Wiretapping and
Electronic Surveillance
in America, 1862–1920

OTHER WORKS BY KERRY SEGRAVE AND FROM MCFARLAND

Beware the Masher: Sexual Harassment in American Public Places, 1880–1930 (2014)
Policewomen: A History, 2d ed. (2014)
Extras of Early Hollywood: A History of the Crowd, 1913–1945 (2013)
Parking Cars in America, 1910–1945: A History (2012)
Begging in America, 1850–1940: The Needy, the Frauds, the Charities and the Law (2011)
Vision Aids in America: A Social History of Eyewear and Sight Correction Since 1900 (2011)
Lynchings of Women in the United States: The Recorded Cases, 1851–1946 (2010)
America Brushes Up: The Use and Marketing of Toothpaste and Toothbrushes in the Twentieth Century (2010)
Film Actors Organize: Union Formation Efforts in America, 1912–1937 (2009)
Parricide in the United States, 1840–1899 (2009)
Actors Organize: A History of Union Formation Efforts in America, 1880–1919 (2008)
Obesity in America, 1850–1939: A History of Social Attitudes and Treatment (2008)
Women and Capital Punishment in America, 1840–1899: Death Sentences and Executions in the United States and Canada (2008)
Women Swindlers in America, 1860–1920 (2007)
Ticket Scalping: An American History, 1850–2005 (2007)
America on Foot: Walking and Pedestrianism in the 20th Century (2006)
Suntanning in 20th Century America (2005)
Endorsements in Advertising: A Social History (2005)
Women and Smoking in America, 1880 to 1950 (2005)
Foreign Films in America: A History (2004)
Lie Detectors: A Social History (2004)
Product Placement in Hollywood Films: A History (2004)
Piracy in the Motion Picture Industry (2003)
Jukeboxes: An American Social History (2002)
Vending Machines: An American Social History (2002)
Age Discrimination by Employers (2001)
Shoplifting: A Social History (2001)
Movies at Home: How Hollywood Came to Television (1999; paperback 2009)
American Television Abroad: Hollywood's Attempt to Dominate World Television (1998; paperback 2013)
Tipping: An American Social History of Gratuities (1998; paperback 2009)
Baldness: A Social History (1996; paperback 2009)
Payola in the Music Industry: A History, 1880–1991 (1994; paperback 2013)
The Sexual Harassment of Women in the Workplace, 1600 to 1993 (1994; paperback 2013)
Women Serial and Mass Murderers: A Worldwide Reference, 1580 through 1990 (1992; paperback 2013)
Drive-in Theaters: A History from Their Inception in 1933 (1992; paperback 2006)

Wiretapping and Electronic Surveillance in America, 1862–1920

KERRY SEGRAVE

McFarland & Company, Inc., Publishers
Jefferson, North Carolina

LIBRARY OF CONGRESS CATALOGUING-IN-PUBLICATION DATA

Segrave, Kerry, 1944–
    Wiretapping and electronic surveillance in America, 1862–1920 / Kerry Segrave.
        p.   cm.
    Includes bibliographical references and index.

    ISBN 978-0-7864-9624-2 (softcover : acid free paper) ♾
    ISBN 978-1-4766-1740-4 (ebook)

    1. Wiretapping—United States—History—20th century.
    2. Wiretapping—United States—History—19th century.
    3. Electronic surveillance—Social aspects—United States.
    4. Privacy, Right of—United States. I. Title.
TK6383.S44 2014
363.25′2—dc23   2014030858

BRITISH LIBRARY CATALOGUING DATA ARE AVAILABLE

© 2014 Kerry Segrave. All rights reserved

No part of this book may be reproduced or transmitted in any form or by any means, electronic or mechanical, including photocopying or recording, or by any information storage and retrieval system, without permission in writing from the publisher.

On the cover: stenographer using dictaphone by Bain News Service on April 27, 1911 © Library of Congress

Printed in the United States of America

McFarland & Company, Inc., Publishers
  Box 611, Jefferson, North Carolina 28640
    www.mcfarlandpub.com

Table of Contents

Preface	1
Introduction	3
1. The Civil War	5
2. Stealing Electricity	12
3. Worries About Privacy	19
4. Wiretapping the Bookies	24
5. Wiretapping the Markets: Stocks, Commodities and Bucket Shops	64
6. Wiretapping Other Businesses	82
7. Wiretapping Labor	98
8. Wiretapping: Other Crimes and Personal Use	107
9. Wiretapping: Politics, Laws and Police	112
10. Dictograph: the First Bug	141
Chapter Notes	193
Bibliography	205
Index	215

Preface

This book looks at wiretapping and electronic surveillance in America over the period 1862 to 1920. During those years the surveillance system moved from nonexistent to a complete system in which everyone in the country was a potential target of surveillance, if somebody had a reason. Initially only telegraph wires were tapped, because nothing else existed to be tapped. Since the private telegraph wire was something found only in large businesses and offices the individual citizen was at no risk. Within a couple of decades the telephone was added to the items that could be tapped. This time the individual citizen was at risk since the telephone quickly found a place in virtually everyone's home. Then in 1910 came the first bug, the dictograph. This allowed room conversations to be surveilled, a complement to the tapping of phone conversations. Everything was covered and all citizens were potential targets, even those few who had no phone. Over that period of time the skill and expertise needed to tap someone's wire or home became easier and easier to acquire. It was difficult to tap telegraph wires as telegraphic skills were needed and not easily learned to the high level of proficiency necessary. With respect to the telephone tap the skill level needed decreased considerably. Finally, with the emergence of the dictograph, the skill level necessary dropped still further. This was the type of gadget that one could buy off the shelf in a store and then go out and use efficiently that same day, although break and enter skills might still be needed.

Research for this book was conducted using online databases. Most useful were the Library of Congress's "Chronicling America" site, newspaperarchive.com, the *New York Times* database, and the *Washington Post* database, as well as others.

Introduction

Over the period covered by this book bugging moved from being a virtually unknown activity to one that affected mainly one set of crooks scamming another set of crooks, to one that involved police and private detective agencies and corporations (often acting in a hazy consort with the distinction of who was doing what), to an activity that potentially targeted everyone and which received a great deal of media attention, especially in the period 1910 to 1920, and particularly in 1912 and 1913.

Chapter 1 looks at the situation during the U.S. Civil War, wherein the tapping of telegraph lines was practiced by both sides in the conflict. The surveillance state we live in today in which everyone is surveilled all the time (at least with respect to online activities) got its start here, in 1861 and 1862, on the battlefields of America.

Stealing electric power though the tapping of those power lines is the subject of Chapter 2. Such tapping, and the tapping of telegraph lines, found the American legal system to be unable to deal with such activity. Theft was always regarded as the taking of something that was physical, tangible. There was a lag as laws had to be written or rewritten in all the states to cover tapping. And those laws were enacted.

Worries about privacy are discussed in Chapter 3 as some opinion makers weighed in on the subject. But their numbers were few and there was no real opposition to the idea and the art of tapping.

Chapter 4 focuses on wiretapping the bookies. One of the reasons that authorities were sometimes slow in going after tappers was that the first wave of wiretappers devoted much of their attention to hitting the bookie joints. For many in positions of authority it was a question of one set of crooks scamming another set of crooks, and why bother to hunt them down or prosecute them?

Wiretapping the stock exchanges, commodity markets and bucket shops is the focus of Chapter 5. Much of this activity also involved establishments (bucket shops) that were illegal or immoral, or both. But legitimate stock brokers were often tapped and that provoked a more vigorous response from officials. Other businesses that were tapped is the subject of Chapter 6. Included in that group are newspaper press agencies.

Chapter 7 looks at the wiretapping of labor, both by labor groups and of labor groups. They were a regular target of such activity from the employers. Chapter 8 looks at other crimes such as using fake wires in an effort to rob banks and the personal use of tapping in an effort, for example, by someone to obtain evidence against a spouse that could be used in a divorce case.

Chapter 9 explores wiretapping with respect to politics, police and the laws. It is here that the interaction between the state and corporations and private detectives can be seen. Even though tapping was illegal by this time that never bothered the state. The New York Police Department began tapping telephones in 1895 and apparently never stopped, in the period covered by this book. By the early 1910s they added the dictograph to their tapping arsenal.

Finally, Chapter 10 looks at the dictograph, its rise from nothing in 1910 to its fascination for the media in that decade, and to its widespread use and seeming acceptance by the end of that decade.

1
The Civil War

Through the 2010 and 2013 revelations by Bradley (later Chelsea) Manning and Edward Snowden, respectively, we all came to realize that, with regard to electronic communications, we were mostly all surveilled most of the time. And it was not just by the state but by corporations as well. It all started 150 years earlier on the battlefields of America during the Civil War when George Ellsworth, a member of the Confederate States of America Army, tapped Yankee telegraph lines. Ellsworth was not the only tapper of the era but he was the one, apparently, most often remembered. It continued in 1895 when the New York Police Department began to tap telephone lines. That fact was not revealed to the public until 20 years later, by which time the NYPD was doing a booming business in telephone tapping. The practice of electronic surveillance really took off in 1910 when the dictograph arrived on the scene and made it easier for the average person to engage in tapping. The dictograph was the first ready-to-use bug, more or less, that anyone could employ—buy it off the shelf and utilize it with little or no skill or training necessary. Politicians dictographed other politicians; corporations dictographed other corporations, as well as labor unions; stock brokers bugged other stock brokers; and the police dictographed everybody. By 1920 we were well on the way to the world that George Orwell described in his dystopian *Nineteen Eighty Four* novel and which, due to Manning's and Snowden's disclosures, has arrived.

At the time the United States Civil War broke out the Morse telegraph was still in its infancy, a little more than two decades old. The connection of the United States by telegraph from west coast to east coast took place in the latter part of 1861. While that development put an end to the legendary Pony Express mail service it opened the door for wiretapping throughout the nation.

One of the earliest reports on the practice of tapping telegraph wires during the Civil War appeared in September 1862, in a newspaper published in Junction City, Kansas. According to the account, CSA commander John Morgan was still riding up and down, apparently pretty much where he chose, in Kentucky and Tennessee. He had last been heard of in Gallatin, Tennessee. His band of men had destroyed the bridges and thus cut the railroad connection north, but, said the account, "not the telegraph; which indicated that they have tapped the wires and are taking the union dispatches from them."[1]

During September 1863 the Yankee forces were establishing themselves in control of Morris Island (an uninhabited island in the Charleston, South Carolina, harbor) by occupying it. A Union reconnoitering party that included a captain, lieutenant, telegraph operator, and soldiers was captured on the Savannah railroad below Charleston. According to a Southern source, "They had tapped our wires and were endeavoring to obtain information, but got none."[2]

Other accounts of Civil War–era wiretapping were retrospective, published mostly in the 1880s. One such article briefly summarized a monograph on the military telegraph by noting that tapping the wires to steal the secrets of the enemy as well as to send misleading dispatches was a frequent tactic by both sides in the Civil War, "but the results of this kind of enterprise do not seem to have been often very large, the military commanders being too cautious." The success of Union general William Tecumseh Sherman's raid into West Virginia in September 1862 was attributed to the information that the general's telegraph operator obtained while holding a Confederate telegraph office, a fact unknown to that side, and taking down their messages.[3]

When an old Civil War telegrapher was interviewed in 1884, he explained that not many messages taken from the wires by the opposing side divulged much in the way of information because most were sent in cipher and were untranslatable without a key. This unnamed veteran said the tapping of telegraph wires was perfectly fair and was practiced extensively by the military telegraphers on both sides during that war. While destroying as much as possible of the CSA property, it was often a rule with the Union commanders to leave their telegraph lines alone. Each side had operators on the alert to take advantage of any opportunities that arose "and used all the means ingenuity would suggest to defeat each other's stealing. The union men, however, learned more than the confederates. The federal army cipher code was so perfect that the enemy, although they captured by their operators many messages in transit over the wires and published them in their newspapers with appeals for translations never could find its key." According to this source a telegraph oper-

CUTTING THE WIRES.

A line drawing from 1888 showing wiretapping of a telegraph line in progress during the Civil War. It was a common practice engaged in by both sides in that conflict.

ator with Union General George Henry Thomas stayed in a tree-top for 48 hours in a howling rainstorm, listening to Confederate dispatches passing over the wires. He declared the most successful instance of wiretapping was by General Robert E. Lee's operator, a man named Gaston, who managed to tap General Ulysses S. Grant's private wire to Fort Monroe.

For three weeks he listened undiscovered to cipher messages, copying and forwarding them to his commander who, of course, could not make them out. But Gaston got one message in plain language—the only one that was sent in all that time. But it was important. It was from the quartermaster in Washington and reported that 2,500 cattle would be landed at Coggins Point. Lee's men were on hand to meet them and got the beef after a fight with the Union troops and feasted on it for 40 days.[4]

According to an 1887 account, telegraph operators in Tennessee and Kentucky were always in danger of being surprised and captured. Although they were non-combatants the Confederates would not consent to treat them as civilians. This article also noted that the tapping of the wires was resorted to by both sides during the war. Anyone caught in the act of wiretapping would, of course, be executed at once as a spy. George Ellsworth, a "noted" operator attached to the command of John Morgan, was "very expert" at the business of tapping. He always carried with him a pocket-sized telegraph operator's instrument which he could attach to the main line. By the aid of this device he could be put into the circuit without opening or breaking the electrical current. During one of Morgan's raids the operator at Gallatin, Tennessee, was captured before he could get out a danger message by wire. Ellsworth took his place and for several hours deceived the other offices along the line. He reported the railroad track to be clear of guerrillas in order that the trains might be sent along, which would of course have been captured. It was said Ellsworth was so accomplished as an operator that he was able to cut a wire, place the two ends against his tongue and by the vibrations read all messages coming over the lines. At one point Union General Joseph Hooker thought there was reason to believe Robert E. Lee was getting too much news. Investigation discovered a wire tap that led them to a small house. The operator at the Union end of the line was discovered and arrested, and not heard of again. It was the custom of raiders from either side to cut down telegraph poles, chop them into stove lengths and twist the wires into knots. Then an operator named O'Reilly attached to Union Major General Benjamin Grierson's command came up with a simple plan that rendered the telegraph lines useless but left them intact and standing. O'Reilly now and then cut the wires, inserted between the ends a non-conducting substance and then welded the ends together. The wires looked as good as new but could not be made to work.[5]

Joseph Orton Kerbey authored a book about his experiences in the Civil War, published in the 1880s. He was under the command of Union Major General George Stoneman and came to be the person in charge of the wiretapping business for that unit. He was once disappointed to find

ELLSWORTH TAPPING A WIRE TO LEARN THE ENEMY'S SECRET.

An 1896 sketch of George Ellsworth busy at what he was considered expert at, wiretapping.

advance men in his group had, in going ahead, cut down nearly a mile of telegraph poles to burn with their railroad ties. They had gathered the wire up and piled it in heaps on the fire. "This was exactly what I did not want done. My purpose was first to tap the wires and attach my pocket instrument and have some fun out of it," wrote Kerbey. He had discovered a safer and surer method of destroying telegraph wires. A telegraph line could be rebuilt and used with the wire lying on trees or even fences, in dry weather. Therefore the cutting out of a mile of poles was not an effective interruption. Kerbey's plan was to take some of the small magnet wire (copper) that was so thin as to be almost invisible, attach that to the insulator-hook or wire at the top of the pole, run that thread of wire down the pole, embedding it if possible in some seam or crack to further conceal it, and at the bottom of the pole run the other point of wire into the ground. Thus, all electric communication was effectively conducted off the channel. All signals would go to the ground and die there. But the wire seemed to work as usual so the source of trouble was hard to find. It could take days, perhaps, to find the problem. Kerbey did his tapping during one of Sherman's raids, probably in 1863. One time he was tapped in and talking to a Confederate operator in, he thought, Richmond. Stoneman himself happened by and participated. Kerbey explained he wasn't sure what to say to the questions posed to him by Richmond. Stoneman told Kerbey to say the Yankees cut the wires and that they had then been driven back home again. Kerbey added that he was a repairman sent out to fix the wire and that the Yankees had been chased back by Confederate General J. E. B. Stuart.[6]

During July 1863 Confederate General John Hunt Morgan was conducting raids through Kentucky, Indiana, and Ohio. "On his raids he was always accompanied by a telegraph operator, who carried a small electric battery," noted a reporter in 1888. "In Union neighborhoods the bold raiders tapped telegraph wires and caused his operator to transmit and receive messages that gave him valuable information." One time during those raids he stopped at Springfield. He had thought he might reach Louisville and capture it. But at Springfield, as was his custom, he had the wires tapped to ascertain what his enemies knew. He thus learned that he was expected by the Yankees at Louisville and they were ready for him. As a result Morgan changed directions and plans.[7]

A different account of Morgan's activities in midsummer 1863, by George Kilmer, published in 1896, stated that Morgan was selected to try and stay the fall of Chattanooga. Morgan had two mounted regiments and two small detachments, also mounted, "but the best card in his hand for the crisis was a bold and skillful telegraph operator named George A.

Ellsworth." Morgan had begun to confuse Union forces on July 8 by tapping wires between Louisville and Nashville. After taking off messages showing exactly what the opposing generals knew about Morgan's band and what they were doing to thwart his raiders, he wired out false information. On July 13 Ellsworth captured a Union telegraph operator and managed to get that man to call the Lexington office and ask for the time of day. That was a ruse to get the operator's style, so he could later imitate it. The Frankfort, Kentucky, office lines were cut and all dispatches received at Midway (where Morgan and his men were). In a short time Lexington asked Midway if it would be safe to run a train between those points. Ellsworth's answer was "All right! Come on. No rebels here." Later on, to gain time and a head start, Ellsworth telegraphed to the Union headquarters at Lexington that the raiders had passed Midway and were going toward Frankfort. Ellsworth continued to send out false information as to Morgan's whereabouts during those summer raids.[8]

2

Stealing Electricity

When that new technology, electricity, began to make its way into American life, one of the first things a few did was to figure out how to steal it by tapping into the lines. All the various ways that people used to wiretap and to eavesdrop in the period covered by this book generated problems for the legal system. It was not ready initially to deal with the subject and, as a result, some of the early wiretappers managed to get away with it. But not for long, as the legal system quickly caught up by writing new laws.

One of the earliest pioneers in the area of stealing power was Russell Clifford, a Shattuck Street jeweler in Lowell, Massachusetts. In July of 1887 he was arrested and charged with what was then a novel crime. It was alleged he had stolen "large quantities" of electricity from the local telephone exchange. The first offense, it was declared, was committed by tapping the wires of the telephone company, and the next was by tampering with the company's meter. For some months prior to the arrest the local managers of the telephone company had been convinced that many parties were using the wires without paying but could not solve the mystery. Finally Clifford was detected some 14 months before his arrest tapping the wires of the company on Shattuck Street and stories spread that he was running an independent telephone company. On a stormy night in July 1887 a female operator for the phone company discovered her wire had been crossed in a strange way and she was talking to people who were not on her company's subscriber list. That fact was reported and investigated. Then it was declared that Clifford had obtained some telephones from a local inventor and rented them out to regular subscribers. Reportedly he had three lines running from his shop to various points in the city and he had a dozen subscribers to his independent phone company. Clifford denied he was running a rival exchange but admitted he had tapped the wires of the telephone company. Nothing more was reported on the Clifford case.[1]

Sometime in the first part of 1892 a St. Louis hardware merchant by the name of Gustave A. Tempel was arrested for tapping a wire of one of the electric light companies in that city and securing his lighting free. He was duly arraigned in court before Judge Claiborne, who would not concede that the offense was petty larceny and the grand jury would not allow that the offense was fraud. As a result Tempel was discharged. An editorial on the subject was upset by the idea that the court held the act alleged against him did not constitute theft. "This seems scarcely just. The charge appears to have been admitted. If the man had obtained the illumination of his house or business place by the secret taking of tallow candles, kerosene oil or gas he would have been held guilty," said the newspaperman.[2]

In the spring of 1895 there was a change in the tenants in the Harris Building on Third Street in San Bernardino, California, with the result that the electric wire connection to the house was shut off. Near the end of June that year the discovery was made that the new tenants, Whitehead and Barrett, had restored the connection themselves and had been using the lights for some two months without paying. The manager of the electric company made an application to the district attorney for warrants for the arrest of Whitehead and Barrett. However, upon examining the statutes the District Attorney stated that the laws of California did not cover the tapping of electric light wires, although the tapping of a gas main was a criminal offense.[3]

A March 1897 news article related that it had been decided by a court in Germany that electricity could not be stolen. An electric light company discovered a man had tapped their lines and was using their power without paying. The question arose in court whether an invisible, intangible material could be carried off or stolen. That court ruled that only a moveable, material object could be stolen. It decided electricity did not come under the definition and the man in question was allowed to go free.[4]

Germany's decision in the case mentioned above provoked an editorial published in a Sacramento, California, newspaper. "In this city many years ago it was decided by a court that a message passing over the electric wires could be stolen by one who lay in wait to hear the ticking of the telegraphic instruments," declared the editor. "We believe the Supreme Court has so held. The Court of last resort in Germany has just decided, however, much to surprise on this side, that electricity cannot be stolen, because it is bodyless." The court held, in that case, the defendant was guilty of neither larceny nor embezzlement. Electricity was bodyless, therefore not considered movable or personal property, in the eyes of that court. As far as this editor was concerned, that German ruling was "absurd." He was confident, though, that we would not be troubled in America by "any such absurd and strained analysis. We have already provided against such in several States, and probably will do so in all."[5]

Three years later an article observed that tapping the wires for the interruption of information of a commercial value had long since been judicially declared to be theft in the United States and elsewhere, although the article stolen was intangible. That item went on to state that "Germany has now put into operation a law punishing the theft of electrical power, which belongs to the same class of indefinable property."[6]

In the spring of 1897 the New Jersey Legislature disposed of a number of bills by the two Houses on a particular day, which then went on to the governor of the state for signing. One of those bills provided for "a fine of $500 or six months' imprisonment on persons convicted of tapping electric wires and stealing the current."[7]

Around the same time the case of Michael J. Hickey, a saloonkeeper in Louisville, Kentucky, came up in the criminal court. He stood before the bench on April 28, 1897, charged with grand larceny. The alleged offense consisted of Hickey's tapping an electric light line and using the current therefrom. That case had been brought on a demurrer involving the point as to whether or not electricity could be stolen. Judge Noble held that electricity was property and decided the case would have to be tried on its merit. Nothing more was reported on this case.[8]

Michael McManus, an electrician, had a penalty of one year in jail or a fine of $1,000 or both staring at him because of what he allegedly did to the Suburban Electric Light Company in Scranton, Pennsylvania. McManus worked as a lineman in that city for the Illuminating Heat and Power Company. In court in January 1898, it was revealed that his employer allowed its employees a certain amount of electricity for free for lighting purposes at their homes. He had a line going into his home but the company had a meter placed in the man's apartment to see he did not exceed the set amount of free electricity provided by his firm. In the same building where McManus lived the Suburban Electric company also had a wire. McManus, it was charged, tapped that line and used its power. That fact was discovered by Suburban when that company ran a test and shut off all of Illuminating's power to that building. When they did that they saw that the lights in McManus' apartment continued to shine. The current supplied to the building by Suburban was then shut off, with the result that McManus's lights all went out. The outcome of the case was not reported.[9]

Dr. A. E. Peck, a dentist in Minneapolis, came up with a scheme in the summer of 1898 for stealing electricity. He was caught and was brought to court, charged with allegedly having stolen electricity, the rightful property of the Minneapolis General Electric Company. Peck had a suite of rooms in the Medical Building on Nicollet Avenue that were powered by electricity, extending to his lights and also to many pieces of equipment he used in his profession. That suite consisted of four rooms with a meter located at the rear of the rooms. Regularly, each month, he went to the power company's office and complained about the amount of his bill. Then, suddenly, the meter registered so little the company got suspicious. They sent two of their employees out to Peck's office to investigate. Those men quickly found a tap such that the current entering the office bypassed the meter and went directly into the front rooms where most of Peck's equipment was located and used. Thus the meter registered only the electricity used in one of the four rooms. The warrant against Peck was sworn out under the new electric law, which was passed by the previous Minnesota Legislature to meet just such circumstances as this case. The penalty for someone found guilty was a fine of no more than $500 or up to five years in the state penitentiary. No more reports were published on this case.[10]

On September 8, 1900, Frank A. Pigeon, an engineer for two large commercial buildings, was arrested on a charge of stealing electricity from the Milwaukee Electric Railway and Lighting Company. The firm said that Pigeon made secret connections with the power wires of the company on Mason Street and used the power therefrom to light the buildings of his employers. Company spokesman Beggs said he thought such wiretapping had been going on for years in other parts of the city and that Pigeon was not the only offender. Beggs said the case could be prosecuted to the fullest extent possible and an example made of Pigeon. Some months earlier Superintendent Rau of the lighting company said he discovered that when the power plants of the two buildings were shut down late at night there would be lights on in both of the buildings, especially when meetings or dances were being held in the buildings. He claimed that an investigation showed that a number of illegal and unauthorized street connections had been made to the company's wires and that the detectives of the firm were actually shown by Pigeon where the connections had been made, underground and overhead, at various places in the buildings. No more was reported on this case.[11]

Ignatz Beck was the proprietor, in December 1901, of a pharmacy located at 1348 Ellis Street in San Francisco that, said a reporter, "is greatly admired by the neighborhood because of its brilliant lighting at night." There were 30 electric lights in the store, which were supplied with power by the Independent Electric Light and Power Company. As Beck's bills were considered small for the quantity of power furnished, an investigation by the company was launched. On July 23, 1901, the suspicion arose that Beck had tapped the wires of the firm. It was determined from the investigation that 20 of the lights used were surreptitiously supplied with current. The power used for those lights was not registered on the meter. The company swore out a complaint against Beck charging him "with willfully and unlawfully connecting a wire with the apparatus of the electric company." That action was taken under a recent addition to the California Penal Code relating to electric power companies, and was said to be the first action of its type ever undertaken.[12]

Beck's trial on the charge of tapping the main conduit wire of the Independent Electric Light and Power Company with the objective of defrauding the firm was held before a jury in Judge Fritz's court on January 2, 1902. The main witness to appear was Seth Cohen, inspector of meters for the company. He examined Beck's premises the previous November and discovered that the firm's wires had been cut and tapped by two wires that fed 38 incandescent lights that were not registered at the meter. Beck's contract with the company called for five incandescent lights and two arc lights. Another company employee corroborated Cohen's evidence. William G. Pennycook, inspector in the city department of electricity, testified to having seen on the previous July 23 about 40 incandescent lights burning in Beck's store. He asked Butler, the clerk in the store, who had done the job and Butler replied, "The boss." Butler was instructed by Pennycook to have Beck call in at his office to see him but Beck never did so. In court Beck admitted he burned 38 incandescent and two arc lights but when shown that his bills never exceeded $11 per month and were as low as $6 he could not explain it. The pharmacist was convicted by the jury which was out for deliberations, said a reporter, "but a short time." It was the first case of the type prosecuted by the electric company.[13] On January 4, 1902, Ignatz Beck appeared for sentencing. Judge Fritz sentenced him to pay a fine of $100 with the alternative of 50 days in the county jail.[14]

H. A. Wood, representing the Independent Electric Light and Power Company in San Francisco, swore out a complaint before Police Judge Fritz on March 27, 1903, for the arrest of Louis Arzner, proprietor of the Ferry Café, located at 16 Market Street, and Carroll Holmes, electrician, on the charge of tapping the electric light wires of the company without authority, a felony under section 499A of the California Penal Code, approved on February 22, 1901. That company furnished electric light power to a jewelry store at 20 Market Street and it had recently discovered that the service wire of that store had been tapped outside the meter and the current conveyed to the rear portion of the Ferry Café. A check meter was put on and it was alleged that 20 kilowatt hours per day were stolen, being of the value of at least $1. Arzner was taking his electric light from the San Francisco Gas and Electric Company and it was also learned that there was a corresponding drop in the bills paid by Arzner to that company after the wires of the Independent were tapped. Two employees of the Independent went to the Ferry Café on March 24. One remained in the café on watch while the other cut off the current stolen from the jewelry store. Immediately, the electric lights in the rear portion of the cafe, 31 in number, went out, regarded as conclusive proof of the theft from the Independent. One of those investigating employees remained in the café. Later, a man with a coil of wire over his shoulder entered the café. It was electrician Holmes. He went to work reconnecting the cut wires.[15] While nothing more was reported on this case, just three weeks later it was reported that the Ferry Café at 16 Market Street had been bought by new owners, Gus H. Kilborn and J. Emmet Hayden.[16]

Late in December of 1905, W. T. Conway was accused of using what Utah Light and Railway Company officials regarded as a "jumper" for the purpose of defrauding the company of current supplied to the Conway home at 374 Fourth Street in Salt Lake City. Arrested on December 28, he was charged with a violation of Section 4,371, Revised Statutes, which made it a misdemeanor to employ a device to secure electricity without causing the current to pass through the meter provided for registering the quantity of electricity consumed. His method worked successfully from about the middle of October until December 9, when the company caught on to the scheme. On December 9 a company official visited Conway's home, supposedly to read and inspect the meter. But when he checked he found that when he turned off the main switch ahead of the meter, the lights continued to burn. More inspection revealed jumper wire hidden by a shelf just above where the wire was tapped. Utah Light and Railway did not reveal what had tipped them off. Conway admitted he did the wiring himself. Under the Utah statute, the offense was a misdemeanor and was punishable by up to six months in the county jail or a fine of $300 or both. On January 3, 1906, Conway was found guilty in court and was fined $25 and costs.[17]

Wiretapping, or the placing of a jumper on an electric company wire, was just plain larceny, according to a decision made by Judge J. M. Bowman in police court in Salt Lake City in October 1909 when he found the Western Macaroni Company guilty of larceny and continued the case for sentence later in the week. That the Western Macaroni Company stole electricity from the Utah Light and Power Company was said to have been proven during the trial of the case. Theft had been continuous since July 12 of that year when, according to the evidence, it was revealed a mechanic called at the company plant and offered to install an apparatus that would save the firm on its electricity costs. The device proved to be a jumper that carried the current past the meter and into the plant without registering it. A trouble call caused an investigation by an employee of the power company and the discovery of the jumper. No report of the disposition of the case was reported.[18]

John Hodgson of South Omaha, Nebraska was convicted in the county court in August 1911 of wiretapping and stealing electricity from the lines of the Omaha Electric Light and Power Company. Punishment levied against Hodgson was a fine of $50 and costs. It was reported that the power company was insisting on the prosecution, according to law, of all who were caught tapping its wires. Some had been apprehended who had arranged a circuit around the meter while others had been caught who had arranged a line connecting their premises with the power line on the street.[19]

Frank Pulver, superintendent of the Wolfe Building in New York City, and John Eisermann, chief engineer of the same building, located at 518–526 West 26th Street, were held on August 15, 1913, by Magistrate Deuel for trial on charges made by the New York Edison Company that they used $30,000 worth of its electricity by tapping a street conduit. An investigation led to the discovery, by Edison investigators, of the tapping. George T. Leach, former night engineer of the Wolfe Building, said in his affidavit that he was instructed by Hugh Clinton, the former day engineer of the building, and by Pulver to connect the building with the Edison wires. Clinton also submitted an affidavit in which he claimed the use of the Edison power without the company's authorization was common knowledge to those in charge of the building. The tapping went on for six months. Harris Wolfe, one of the heads of the building, said his employees tapped the wires on their own initiative and not on his orders or instructions. He also argued that the amount of electricity said to have been stolen was exaggerated and that no more than $60 worth of power could have been taken during those six months. No outcome was reported.[20]

3

Worries About Privacy

The arrival of the telephone on the American scene caused at least a few observers of the time to worry about what it would mean for privacy, both for the individual and for institutions. No such worries or speculations had greeted the earlier arrival of the telegraph on the scene. It was, of course, a technological improvement that was never meant to be a part of the average American's home and daily life. It was designed for the use of businesses and various other types of large concerns. Individual Americans usually only used the telegraph service very infrequently and when they did they went down to the Western Union office in person to send their wire. It became a contrivance in many a Hollywood film for a telegram to be delivered to someone's home—it was dramatic moment because the receipt of that rarely encountered wire always meant bad news. It was very much different with the telephone as that technological advance was indeed meant to find a place in every person's home and to be used on a regular basis, soon to become daily use. It was no easier in the late 1800s to tap a telephone wire than it was to tap a telegraph wire but the second skill that was involved in tapping telegraph wires, that of being able to read and send and receive messages in Morse code, was absent from phone tapping. It was easier by one level.

An editorial on the subject appeared in a New York City newspaper in March 1877. That editor was hopeful the telephone would be a more secure communications medium than the telegraph. "As things are now the most confidential communications between distant points may be tapped at any office in the circuit," he wrote. "Doubtless many of our readers who have stopped while waiting for a train, to have a chat with the telegraph clerk at a way station, have heard him mention that his instrument, ticking away in lively style, was repeating a message in which he had no concern—a message between other places on the line. In war times a lively business used to be done by tapping the wires." The editor hoped the arrival of the telephone would dispense with the need of taking all the telegraph operators on a circuit into one's confidence, narrowing the matter down to two, the sending and receiving operators (as the early telephones had and required, even for local calls). As well, he hoped the phone might dispense with the telegraph clerk altogether and enable the sender of a message to talk into the very ear of the receiver of the message. "But then suppose, in the latter case, that somebody who has no business in the affair applies his telephonic funnel somewhere along the line while a very confidential message is passing," he worried. "It is yet too soon to predict whether the new invention will fully secure what is most of all needed, the sacred privacy of telegrams."[1]

A second editorial on the telephone and the subject of privacy appeared in January of 1878, originally in a newspaper from Toronto, Canada, but reprinted in American papers. At that early date in the telephone's evolution the editorial wondered if the phone was already a failure. Noted was the fact the telephone had been accepted by the public at once with little popular or scientific hostility to contend with. As soon as it was announced that sound was transmissible by electricity "it was embraced." The editor went on to observe, "Another serious defect in the telephone is the ease with which the wires can be tapped. Hitherto no one but an experienced telegrapher could steal lightning. To tap a wire that is running under the Morse or Wheatstone code of signals requires knowledge of those signals and the possession of instruments" but the telephone was much easier to deal with, from the perspective of a potential wiretapper. "The invention is thus entirely untrustworthy in war, in which business it was hoped it would find its chief value. Any traitor could tap the wires at any point and convey to the enemy what he learned," he grumbled. "It would not be necessary for him to cut the wire or put it out of circuit, and the operators at either end could not detect that their line was being milked." Thus, this editor concluded, at present it seemed to him that the practical uses of the telephone were limited to those rather rare instances where a completely isolated line could be secured and to situations where the liability to stealing the messages en route would not be a fatal objection—"an objection, by the way, which is likely to prevent the general use of the telephone for press news."[2]

Improvements were announced from time to time with respect to the telephone and telegraph, most of which never came to pass. One thing they usually had in common was a claim that this particular invention or improvement would help do away with wiretapping, or at least make it more difficult. A brief report on the heliograph appeared in 1879. It was used for the first time in war by the British troops that operated against the Afridis (a Pashtun tribe based in present-day Pakistan). That instrument consisted of a circular mirror moving upon a universal joint and supported on a tripod. Wishing to send messages, the operator, by a quick elevation or depression of the glass so as to catch the full glare of the sun, was able to throw flashes a distance of 25 miles. The length of the flashes corresponded with that of the flashes in the Morse alphabet, by which the messages were translated. According to the news account, "This method of field telegraphy does away entirely with the dangers of wire tapping and cutting by the enemy."[3]

A somewhat humorous editorial on the telephone appeared in the summer of 1880. In this piece it was predicted that phone usage would soon sweep the nation, with the day approaching when it would seem quite as natural to conduct business transactions, arrange affairs of state, examine witnesses, have social chats, sing songs, and make love through that agency, as is was to do all those things face to face. "But the peril to statesmen, financiers, rogues and lovers will be so increased by reason of the rare facilities for discovering secrets afforded by tapping the wires, that mysterious forms of speech will become as common as cryptograms in telegraphic communications. Never until then will the nation be secure," concluded the editor.[4]

Another touted improvement was profiled in a March 1883 story. The National Secret Telephone Company was said to have a secret system to enable the transmission of messages between the sender and the receiver in such a manner that they could not be taken off from the wire by induction or by cutting in and tapping the wire. To effect that the impulses from the transmitter were divided and transmitted over two independent lines of wire, one-half passing over one circuit and one-half over the other. At the receiver end those two halves were gathered up, adjusted into their proper order and from a single wire articulated in distinct words. Thus, the word Constantinople was sent in code, using this system, as follows; the sounds C, N, A, T, O, and E were sent over one circuit, while the sounds from the speaker for O, ST, N, IN, and PL were sent over the second circuit. To the ear of someone tapping a single line the message sounded like nothing more than a meaningless buzz. In conclusion, wrote a journalist, "in practice it is not expected that the secret system will be generally employed except by brokers and others whose business is necessarily of a private nature, and the double lines will be put in for it only when they are expressly desired."[5]

A number of journalists, scientists, bankers, and the like were invited in April 1886 to view a practical test of the printing telegraph by the International Printing Telegraph Company. This device, it was claimed, would largely supplant the telephone. It looked like a typewriter, sort of, and while it printed one or more copies of the message before the operator's eyes, it did the same on the other end of the telegraph wire, hundreds or thousands of miles away. One advantage was that the sender got a copy of his message. Supposedly this device was completely accurate, not subject to operator error. Another advantage that "it is impossible to read by sound or to intercept a message by tapping the wires, thus rendering secrecy complete."[6]

Early in 1898 a revolution was announced in the field of telegraphy for the use of businessmen "and to some extent take the place of the telephone." It was a machine for printing as well as transmitting telegrams. It delivered a message whether the receiver was in his office or not. "Perfect secrecy is attainable by the telescriptor and that is not true in the case of the telephone," enthused the article. Telephone operators could always listen in. "So, too, confidential correspondence by telegraph is practically impossible without resort to bothersome ciphers," concluded the piece. The telescriptor was, it was said, "impervious to tapping."[7]

Professor Henry T. Rowland of Johns Hopkins University was reported to be ready to test his multiplex telegraph apparatus in Baltimore, at the very end of 1898. It was designed to send several messages to be sent and received at the same time over the same telegraph wire. Designed to handle as many as eight messages at a time, a reporter declared of it, "It renders wire-tapping an impossibility."[8]

Wireless telegraphy was touted late in 1907. One disadvantage of this new system was that messages were being picked up by ships and land stations for which the wireless messages were not intended. And, wrote the newsman, "so generally is it known that at present there is no privacy in Hertz wave communication that few persons, if any, resort to it for business of a commercial nature. When tuning systems come into more general use greater secrecy will doubtless be feasible." He added that "even then the practice of tapping a wire will probably be paralleled if a sufficient inducement for intrusion and theft is afforded. Instruments have been invented by means of which an electrician can ascertain the frequency of the waves emitted by a tuned transmitter; the information thus obtained enabling him to adjust his receiver so that it will respond to those impulses. This apparatus could, of course, be put to an illegitimate use, with an excellent chance of avoiding detection. There would apparently be no clue by which the perpetrator of the offence could be found, whatever suspicions might be entertained." He added, in conclusion, that stealing messages would, of course, do a thief no good if the sender employed a suitably private code.[9]

3. Worries About Privacy

In the spring of 1914 a man named George Stewart Smith of New York City was circulating petitions to the Public Service Commission of the Second District of New York asking it to investigate the possibility of obtaining a device that would ensure when a telephone call took place it would be protected from eavesdroppers. He thought such devices to protect privacy existed. One catalyst for this, as he noted in his petitions, was the revelations made public during the stormy impeachment procedure against New York Governor William Sulzer. (See Chapter 9.) His petition was called "A Petition for Telephone Privacy." A spokesman for the Commission said that if enough interest were shown it would "undoubtedly" cause an investigation to be made. Union N. Bethel, president of the New York Telephone Company, was asked if he knew of any devices in existence that would produce privacy. He replied, "I don't know of any which would prevent a person, who was determined to do so, from tapping a wire and hearing what was going on." And that was even if those telephone operators were eliminated. He did admit that telephone operators could listen in to conversations, under the current system. But, he added, the amount of work each operator had to do, along with the constant supervision the operator was under from the monitoring system used by the company, made it a practical impossibility.[10]

4
Wiretapping the Bookies

The most active users of the technology to tap telegraph wires in this period were, by far, those who set out to beat the bookies out of some of their cash. Since bookie joints (called pool rooms in this era) were mostly illegal, most of the time, in most places, the law was very ambivalent about pursuing complaints in this area. Officials viewed the whole affair as one set of crooks scamming another set of crooks. Thus many, if not most of these scams, when perpetrated, were not reported to the police at all. Journalists got wind of these stories and published them. One group that did pursue such complaints was the telegraph companies. The Western Union Telegraph Company was the largest such concern and the money received from bookie joints for supplying them with more or less "instant" race results was said to be very lucrative. Often the authorities put pressure on the telegraph companies to not supply the bookies with telegraph services. On the other hand, those companies came under the common carrier concept. People usually associated the concept with the transportation of goods and passengers by railroad companies and the like but it extended to other areas. Under common carrier doctrine a company had to provide the same service to all who applied for that service and to charge the same rate for that class of service to all who applied. Thus, it was difficult for the police to demand that Western Union stop servicing the pool rooms: the company fell back on the common carrier idea—that is, it had to provide the service.

One of the earliest reports of this new type of criminal activity appeared in a newspaper on July 8, 1883. It told the story of the tapping of the wires between the racetrack in Long Branch, New Jersey, and Philadelphia a little earlier and the sending of a false report by which Philadelphia bookies were swindled out of several thousand dollars. According to one telegraph man the reporter interviewed, "Such attempts, in one way or another, are made every year.... A wire is so easily tapped that the companies are wholly unable to furnish absolute protection." Reportedly, an attempt to beat the Long Island City pool rooms during the previous summer was discovered just in the nick of time. It was during the Saratoga races and the intention was to get the race results ahead of the bookies. Western Union had an office in Saratoga and one in Long Island City. Scammers determined to tap Mutual Union news and run it around through an office in New York City and then on to Long Island City. Accordingly, a Mutual Union lineman was bribed and he tapped the wire right in the cupola over the Mutual's office at 137 Broadway. Dispatches from Saratoga thus passed through this "new" office before they reached the Mutual Union office. From their uptown office the gang sent the message by a private wire to Long Island City and saved several minutes of time. That is, they would have done so if the Mutual Union people had not discovered the game a few minutes before 1:00 p.m. when the races were called. How were the gang able to pick out the right wires among the hundreds strung through the streets?, wondered the reporter. The answer was that there were plenty of old linemen who knew all about the telegraph line of the city. Some of those linemen were always out of work, and it was easy to hire them for such work. In the more rural areas, as on the Long Branch and Philadelphia lines, there were only two or three wires. If a tapper did not get the correct one on the first attempt it was easy to try again.[1]

That reporter also wondered if it was more difficult to intercept a message and send a false one than to get the news in advance. "Not very much more difficult," a source told the newsman. The wire was cut and a wire spliced to each end. Those wires were led to separate instruments in one room. A telegraph operator sat at each one. Messages were allowed to pass through until word was received that the particular horse they wished to bet on was at the start. Then the circuit between the two points was cut and the message from the racetrack stopped with the operator who had cut the line. He answered for the point to which the message was supposed to go, and the operator at his elbow, using the signature of the operator at the track, sent on the false message. Then the cut wire was spliced and all traces of the job were removed. "It would take a smart man to find the place where the cut was made. It could be discovered only by accident, because they are so many spliced breaks in every line," explained the telegraph man to the journalist. Next the reporter asked his source if there was any way to protect the bookies. "If they would take their news in cipher they would be perfectly safe. Or they could have their messages repeated. They will not adopt either method. The trouble is they want to save every second of time." As a last question the newsman asked the source if the scammers usually succeeded. The answer was that they usually did not, easy as it was to tap the lines. It was then four years since Kelly and Bliss (a bookie concern) were so badly beaten in Long Island City by a false dispatch, but "[a]n occasional success makes up for the failures, and therefore the wires are cut with each recurring season."[2]

With respect to the above cited Long Branch to Philadelphia scam, it was reported that Superintendent W. C. Humstone of the Western Union Telegraph Company was then engaged in an investigation of the "rumors" that the telegraph wires from Long Branch were tapped on July 4 and that false reports of the winner of the Ocean Stakes at Monmouth Park were sent to Philadelphia. Reportedly, the pool room owners in Philadelphia paid out thousands of dollars based on results from the bogus dispatch before they learned that the race had been won by a different horse. It was said that no complaint had been made by the Philadelphia bookies to Western Union. Humstone assumed the bookies were investigating the matter themselves. A brief interruption in the smooth working of Long Branch messages aroused suspicion within the telegraph company, even before changes were made, that swindles had been perpetrated. "Careful" examination of the telegraph lines by Western Union failed to detect anything. The scam, if it was done, said Humstone, was done by expert telegraphers and hence "we suspect that some of our men may possibly have been concerned in it, but so far we have made no progress toward the discovery of the criminals."[3]

Stories about beating the bookies, and attempts to beat them, became a fairly popular topic in American newspapers in 1883. Later in 1883 a lengthy piece was published that was designed to explain to the public just how a wire was tapped. This anonymous reporter asked a Mr. Moore, described as an experienced telegraph operator, how one tapped a wire. Moore's answer was that an expert was needed to be able to pick out the one message wanted from a hundred, stop it, and let the others go on to their destinations. He said a good many people supposed that it was only necessary to cut the wire, attach another to the cut end where the message was expected and run it off whenever it was wanted. But, he continued, the instant such a thing was done the legitimate operator knew it because "My line is grounded." That triggered a call to a repairman who went off to the area in question. The current was broken and the current ran off into the ground and although the tapper could get the message through his instrument if he happened by one chance in a million to hit it the second it passed, the interference would be instantly detected. Well, said the reporter, how did one proceed? According to Moore, it depended whether the potential tapper merely wished to hear the message or whether he wanted to intercept it and send another one in its place. "In either case the great thing is to maintain the circuit," the course pursued by the current over the wire to its destination and back through the ground to the place it came from. In the early days of the science it was thought necessary to maintain it by a wire back to the starting point. Since then they had discovered that the current would pass back through the ground like a homing pigeon through the air.[4]

Moore continued by stating that in order to merely overhear the message it was necessary to make a loop or V running from the wire to the instrument of the tapper and then back to the line. The wire was cut, the loop line attached and run out of the instrument, where the operator was, and then connected with the main wire. It was simply adding another instrument to those in use on the line, for every message was heard at every station on the route. Except for the purpose of concealment, the interloper might just as well attach his instrument to the main wire at the poles. "But he can only overhear; he cannot intercept messages. At the most he can change one word—for instance, the name of a winning horse." Then, asked the reporter, how could a message be intercepted and suppressed? Moore explained that the tapper had to cut the connection in the manner explained, but instead of leading it back to the wire, he stopped it short. For this the tapper had to have two batteries and at least two operators. In such a case the wire could be cut and a line led off from the New York end to one of the batteries; another was led from the Long Branch end to the other battery. Connection was thus utterly interrupted from New York to Long Branch but the circuit was maintained from each place by those batteries. Then the operator at the cut end of the Long Branch line simply constituted himself the New York terminal. He got every message and gave it to his co-operator at the other battery and instrument, who transmitted it to New York. That continued until the expected message. When that came the operator intercepted it, wrote another one to keep the numbers straight—messages went on with a sequential number attached—and his fellow conspirator telegraphed the false message to New York. In the meantime the Long Branch tapper could answer all questions from the legitimate operator who supposed him to be New York and thus averted suspicion. While he was playing New York to Long Branch, his confederate was acting Long Branch to New York. As a final question, the reporter asked Moore what prevented telegraph wiretapping from becoming more common. Because, explained Moore, the important lines were duplex. That is, one wire could carry more than one message at the same time. With an ordinary tapper's instrument those messages, when received, would be unintelligibly mixed. To properly take the messages from a duplex machine an amount of expense was required that would make too large an outlay for such a risky venture.[5]

The last day of the fall meeting of the American Jockey Club, which closed the racing season in that part of the country, was held in the middle of October 1883 at Jerome Park in Fordham, Westchester County (now the Bronx). The wire of the Western Union Telegraph Company, which alone had a line to Jerome Park, was tapped and false messages substituted for the genuine ones. The fraud was not discovered until the return to the city of persons who had witnessed the racing. Long before that, bookies all over the country had paid out the tickets on what they supposed were the winning horses. Amount of the loss was not ascertained but the Associated Press estimated it would not fall short of $75,000 and could easily exceed that sum. It was learned that pool rooms at Boston, Philadelphia and Pittsburgh had sustained large losses from the fraud. As well, it was thought losses would also be great at Chicago, Louisville, and St. Louis, where the betting was heavy. In the second race that day Eelat was announced as the winner in the false dispatch while the actual winner, Britomarie, was listed as finishing third. From that time until the end of the races that day there was a steady succession of false dispatches. It was stated that while tapping of wires had been attempted frequently, never had it met with the success it had that day from Jerome Park. This account said that it had been tried earlier that year on a small scale between Monmouth Park and Philadelphia but without success. A somewhat elaborate attempt to defraud the pool rooms at Hunters Point, New York, had been made in 1882 but was detected accidentally just as the scheme was being put into motion. An employee of Mutual Union Telegraph Company confessed complicity in the plan. With respect to the Jerome Park scam, this report offered the opinion that it was doubtful the criminals would be caught.[6]

A day later it was said that the Jerome Park losses were nearly $100,000. One firm that escaped without any loss was the pool room of Reber and Gallagher in Salt Lake City. A man had dropped into their bookie joint on the day of the scam, Saturday, October 13, and wanted to make a large bet but when he found the firm only paid off bets on the next day the visitor declined to place the bet, and quickly walked away.[7]

The Jerome Park scam attracted a lot of media attention and publicity. An editor for a New York City newspaper remarked that the success of the swindle might well call people's attentions to the "unthinking confidence" that the public had in telegrams. How easy it was to send a telegram saying anything. A telegraph operator taking a wire from someone did not know if it was true or not. In an Alabama case the editor cited a scammer who knew a rich man's nephew was away traveling. He sent the father a telegram, in his nephew's name, saying he had lost his tickets, money, and so on. The uncle replied promptly by sending him a telegraphic money order. The company paid the money to the swindler, not knowing about the fraud. When the uncle sued the telegraph company the court declared it was not the duty of telegraphers to ascertain the veracity of their wires. Whoever received a telegram had the duty to ascertain the truthfulness of it. With respect to the Jerome Park swindle, this editor thought the telegraph company was not responsible as long as it took reasonable care and precautions to keep its lines in good order, and to take prompt precautions when there was any reason to suspect interference.[8]

Also in October 1883, it was reported that pool rooms in Coney Island, New York, had been losing money mysteriously for about four weeks. It was finally noticed that a young man named Harry Fust would enter the pool room of Al Burtis just before the betting was to be closed for a particular race, make bets and invariably win. Confederates of his would operate in the same way in other bookie joints on the island. At last an employee of Burtis reported to his boss that he had been approached by Fust, who said to him, "You can make lots of money if you will furnish me with the Burtis telegraph cipher. We can get along without it, but we want it." That led to the discovery of telegraph instruments in the Iroquois House hotel a few days later. They were used to tap the wire that conveyed the results of races from all over the country to the pool rooms. While the instruments were seized, Fust and his friends were not found. A report a few days later briefly said that two arrests were made of unnamed persons in the Coney Island case and that they had offered a skilled telegraph operator $500 per day for his services.[9]

Bookmakers in Washington, D.C., were reported to have lost thousands of dollars on October 18, 1888, by having their telegraph wires tapped. Since the United States Congress "drove" the bookies outside of the city of Washington proper, the pool rooms had relied upon one wire to bring them all the racing results. Conspirators rented a house and had, wrote a journalist, "no trouble in tapping the wire."[10]

An account in a different newspaper also noted the bookie joints were all on one wire or they could not have been so easily swindled. A single wire connected them all with the Guttenburg track, in North Bergen, New Jersey. That wire was tapped on the 18th during the last Guttenburg race of the day. Just before the race was called there was a rush by bettors for one particular horse. It was some minutes after the time when the horses should have been announced as at the post before the telegraph instruments began to tick out that message. Then there was another long delay before the operator's "They're off" announced the start. In the interval a man rushed into Lynn & Wall's and bet $35 on the suddenly popular horse. Then another man entered the place hastily and bet $100 on the same horse. That was enough to arouse suspicions in Lynn & Wall. An employee there queried the operator, who asked the operator at the New York end of the wire the same question. The answer was ticked back that the delay was the result of bad starts. That reply allayed all suspicions and several more bets of $25, $10, and $5 were placed on the suddenly popular horse. After the operator had ticked "They're off" it was almost immediately after that the instrument declared the popular horse had won. The gap between start and finish was less time than would have been required to run the shortest kind of race. The operator hurriedly asked the New York operator how it all happened and the answer received was that the race had been over for more than 10 minutes. That was considered conclusive evidence that the wire had been tampered with and the result of the race delayed but, said the journalist, "the bookmakers could do nothing but pay the tickets as they were presented." In this case the winner had actually been the suddenly popular horse. No change in the name of the winning horse had occurred: the result had simply been held back from the bookies for a period of time to allow the confederates to place their bets. According to this account all that was necessary was to cut the wire between the track and the pool room and attach an ordinary set of telegraph instruments to each end. When the results came over the wire they were held up and confederates alerted so they could bet on the horse they knew had won, and then the results were sent on. If it worked perfectly no one knew the scam had even taken place, perhaps just a few suspicions raised. The mistake in this case lay in sending the results on too fast: that is, the delay period was not long enough. It was speculated the same thing had been done several times that summer but never discovered. The operator at Lynn & Wall said he noticed the last race was telegraphed by a different man than he usually received the results from—each telegraph operator's style of sending was said to be as different as handwriting in such things as touch, pacing, and so on—but he thought the New York man had simply been relieved. Lynn & Wall said they lost $500 that day from the scam; Jones & Company lost $150 and Miller and Company was scammed for about $200. The night manager for the Western Union Telegraph Company told a reporter he had not heard anything about the wire being tapped, but the bookies were declaring that they were sure it was done.[11]

Still another account on the Washington scam put the loss at $1,000 in total. It also stated that such a thing as tapping the wires had never before been attempted in Washington but in other cities several such attempts had been made. Bookies paid out on the suspect tickets because there was no hard evidence despite the fact they were all convinced they had been taken. A group of those bookies set out after the last race that day to investigate the situation themselves. They discovered a tap and traced the wire to an unoccupied house that had been rented a few days earlier by two men. That house was kept under observation with several of the bookies, or their agents, staying in the house all night. No one appeared at the house until about 5:00 a.m. on October 19, when a lone man arrived. He had come back to retrieve his telegraph instruments and batteries. The police were called and the man, giving the name of Edward Collins, was arrested. Collins was taken to the police station as a "suspicious person," but as there was no law in the jurisdiction covering such an offense, he was released. According to the reporter, the men connected with the fraud were "known" and would be arrested if found. In such an event they would be charged with malicious mischief. There was then no law against tampering with telegraph wires unless the wires were owned or controlled by the United States. Penalty for malicious mischief was, at the discretion of the court, not to exceed one year in jail.[12]

Several days later no arrests had been made. However, a reporter had studied the situation more closely and declared there was no law on the books to punish such an offense, not even malicious mischief. It also said the wires connecting the pool rooms with the racetracks had been tapped several times that season. The bookies had been warned some time earlier that their wire was being tampered with but they did not take any action in the matter, and the result was that they had lost a large amount of money.[13]

A July 1889 account related that a swindle on the proprietors of the Columbia Pool Room on Third Street in St. Paul, Minnesota, had cost them perhaps hundreds of dollars over the previous week or ten days. It all came to light on July 15 when the police arrested six men. The special Western Union wire from the firm's office ran over the roof of the Merchants building to the Columbia. About a week earlier two young men rented room 30 on the fourth floor of the Merchants building and tapped the Western Union wire from the roof of that building. Since that time they had taken off the reports of the races before the pool room received them. After getting the results the winner was written down on a card and dropped into the alley where a confederate grabbed it and ran across the street to the Columbia Pool Room, betting heavily on the winning horse. There were six men connected with the scheme. After a time the bookies began to notice the frequency with which they were caught hard just before the ticker gave up the race results. Then an examination of the wire led to the discovery of the tapping and the scheme. Those six men arrested were all young and gave their names as Edward Kilroy, George A. Whitney, Fred Ward, F. Durham, F. Fisher, and Frank Mitchell.[14]

Prosecution of the above six took place in municipal court in St. Paul on July 17. In this account, remarkably different from the one above, it was said that the lineman squealed while he was engaged in tapping the wire and informed Superintendent McMichael of the scheme. He was told to keep on with the work and a trap was laid for the conspirators, who were arrested before they had the opportunity to beat the pool room out of a dollar. The men now identified as Edward Kilroy, George W. Wilsey and Frank Dunn pled guilty and were fined $25. Frank Ward and Frank Dunham stood trial. Ward was acquitted but Dunham was found guilty, as it appeared that he had furnished the money for the instruments and apparatus for tapping the line. He was fined $75. Reportedly Dunham had spent a total of $6.08 to buy the tapping equipment.[15]

An account published in July 1889 stated that telegraph operators had many opportunities for making money because of the growing demand for wiretapping. It was said there had been cases where operators had engineered schemes that promised them fortunes, and had used the telegraph to carry them out. "As a rule the sufferers [of the wiretapping schemes] have been gamblers, policy men and pool room keepers especially," the journalist explained. One example cited was a scam worked in 1882 by some operators in disfavor with the Western Union firm. It was at the time the pool room men were banished at Hoboken, New Jersey. At that time the race results were delivered by messengers from the nearest Western Union office and not received currently by operators in the pool rooms. The men involved in the scheme rented a house across the street from where several pool rooms were located, purchased telegraph instruments, and strung a wire to tap the line running to the telegraph office from the bookie joints. Two men loitered around the pool rooms and occasionally made bets on horses that had no possibility of winning. Later they were relieved by two other men, always keeping their eyes on the rented house. Every now and then the shutter in one window of that house gave a violent shake. The shaking of the fourth slat of the shutter meant the number four horse had won the previous race, and so on. The loiterers then swiftly went inside the pool rooms to bet in the 15 minutes or so before the tappers sent the delayed message on to the pool rooms. In the two weeks the men worked the scheme it was reported they netted $20,000. This account also mentioned the Jerome Park scam of 1883. According to this account the Western Union firm came down hard on some of its operators. Arrests were threatened. Several operators who were thought to know more than they wished to tell about the crime were suspended by Western Union and shadowed by that company, but nothing ever came out of it. The same type of scheme was planned for Jerome Park in 1888 and perhaps would have come to pass had it not been for the fright of the lineman employed to tap the wires. A house had been rented and instruments and batteries put in, a cable had been run underground and a groove had been cut in the pole where the wires were to be tapped so that the connecting wires could not be seen, at which point the lineman became afraid. On the morning of the day that the false returns were to be sent in he tore up the cable and went back to New York. Reportedly, this was also a big scheme, with men having been sent to St. Louis, Washington, Baltimore, New Orleans, Memphis, and Troy, New York, all of whom had plenty of money with which to bet. As well, a dozen men with money were waiting in New York City for the word to place bets on the horses agreed upon to be sent on as winners. The six operators involved expected to clear $10,000 each and the men behind them would have made that much more.[16]

Wiretappers struck in New York City on November 17, 1889. The wire led to Gleason's Custom House, a pool room in the Bowery that was a few doors north of Grand Street, exactly in the rear of the house in which the tappers plied their trade. Sitting in their room, the tappers let the race results go by untouched until reaching the race they wanted. When that passed through they removed it from the wire, taking off the name of the winning horse and substituting in as the winner a long shot. Before the fraud could be discovered, the men involved cashed in the bets they had made. Then they skipped out of their recently rented room, taking their telegraph instruments with them but leaving behind the rest of their equipment, consisting of 30 brand-new battery cells. It was unknown, said the reporter, how much money Gleason lost, but "The trick which was thus played on Gleason was the same by which pool sellers throughout the city were swindled of thousands of dollars some weeks ago."[17]

A scheme brought to light by what was called "pure accident" in February 1890 was said to show how every pool room in San Francisco (and there were several) had been systematically robbed over the past three months of money said to run into the thousands. Chief of Detectives Lees of the San Francisco police explained how it all happened. On the top floor of the Safe Deposit Building on the corner of California and Montgomery Streets a small room was rented by the plotters, none of whom was at that time arrested. On Saturday, February 15, a Western Union inspector had business with the large network of wires that went over the top of that building. While doing his work he discovered a small wire that had no business being there. At the same time he found that one end of the wire was attached to a special wire maintained for the pool rooms, over which they got telegraphic information about various races on the Eastern tracks. The inspector traced the other end of the wire through a hole in the roof of the Safe Deposit Building. His superiors were informed and the police called in. They found a room with a complete telegraphic outfit. Bookies were said to be elated as they had lost no less than $50,000 in the previous three months. Speculation was that operators in that rented room held back results and passed that information to confederates who hastened to pool rooms where they bought tickets on the winners. Fifteen minutes was about the amount of time the results were held back by the tappers. Suspicion that something was wrong had been aroused in the minds of bookies C. H. Kingsley, S. B. Whitehead, Henry Schwartz, and others because of the large number of winning tickets that were sold just before the results of the race came in over the wire from the track. Complaints were made to Western Union, which concluded that someone in its employ was divulging information to outsiders, or else its wires were being tapped. An investigation by the telegraph company was launched but without any result, until the discovery was made by chance by that inspector. Western Union used its own investigators to probe the situation and also engaged private detectives from a firm run by Harry Morse to solve the mystery. Several individuals who frequented the pool rooms were regarded with suspicion and they were watched. No result was obtained from the investigations. The California Penal Code made wiretapping a misdemeanor, with the maximum punishment being a fine of $100 and/or six months' imprisonment.[18]

Ten pool room proprietors in St. Louis found themselves tricked out of a sum of money estimated to be from $5,000 to $7,000 in May 1890, by what a reporter called "the old wire-tapping scheme." Such a flood of money came in on a certain horse in one of the races that many of the pool rooms closed down the betting on that race well before the usual closing of bets, when the "they're off" announcement came over the wire. Despite that response a great deal of money was still paid out.[19]

Thomas H. Vines, a Western Union operator, was arrested on Monday, June 2, for alleged complicity in the St. Louis scam. He was "sweated" by the police for 24 hours and then released. Finding himself discharged by Western Union, he got drunk. On the evening of June 3 he was seen sitting in his room on the fourth floor of a hotel on Olive Street, before he fell or jumped out of his window to his death. Vines was also known by the names of Costello and T. H. Kane.[20]

William Fallon, described as a "well-known sporting man," and J. W.

Nagle, a telegraph operator who had been discharged by Western Union for alleged "crooked work" with regard to the firm's messages, were arrested in San Francisco on June 23, 1890, and locked up in city prison. Shortly before 5:00 a.m. that day they were caught by detectives in a room on the top floor of 318 Pine Street while preparing for a day's work of wiretapping. In the room was a complete set of batteries, keys, sounders, receivers, and various other instruments employed in telegraphy. They were to be charged with a felony. The California Penal Code, according to this account, provided that anyone convicted of tapping the telegraph wires could be punished by imprisonment in the State Penitentiary for a term not to exceed five years, by imprisonment in the county jail for a term not to exceed one year, or by a fine not to exceed $5,000. Fallon was a member of the well-known San Jose family that also included brother Al Fallon (horseman) and sister Anita Fallon (actress). William had been left a fortune by his father not long before but had spent his money foolishly in the pool rooms of San Francisco. Nagle was said to be an expert telegraph operator. Some two years earlier the managers of local pool rooms complained that the boys who carried the messages containing the names of winners were bribed by a gang to give them the news before it to the pool rooms. To meet that problem special wires were laid directly to the betting offices by Western Union. Those wires were tapped and the bookies were taken to the tune of thousands of dollars. Western Union then connected the pool rooms with the main office by what was known as a "metallic circuit" and after that the tappers were not successful because the telegraph company's galvanometer was sensitive enough to register the change in current caused by tapping or other interference with the resistance of the circuit. There was no tapping for a time. But then, some weeks later, a suspicious trembling of the galvanometer needle indicated someone was tampering with the wires. Police were notified and an investigation was started. Nagle and Fallon were suspected from the start. When they rented Room 9 at 318 Pine Street a few days before their arrest, for the supposed purpose of starting an electric-motor agency, Captain Lees set a watch on the place. The suspected men went to that room at 4:30 a.m. on June 23 and were arrested at 5:00 a.m. By skill they had completed a successful tap, despite Western Union's technological efforts to thwart tappers, and were ready to go. It was said they had invented a tapping instrument that would have offset the galvanometer at the company's headquarters. When the race results from Western Union went out over the special wire to the pool rooms the resistance of the circuit was very carefully measured and the galvanometer showed the "most trifling" change in the resistance caused by a break in the circuit.[21]

A jury tried William Fallon in Judge Murphy's court in San Francisco on August 27, 1890, and lost very little time in returning a verdict of guilty as charged. When he was put on the stand Fallon made no denial of the wiretapping. His only justification was that he did it to get even with the pool rooms, which he claimed, were conducting an altogether one-sided game. In court there was an argument over the exact meaning of the law and its interpretation. That statute read, "Any person who fraudulently reads, or attempts to read, any message, or learn the content thereof, whilst the same is being sent over any telegraph line" was guilty of a felony. Fallon's lawyer argued no evidence was produced to show Fallon and Nagle tapped the wire "whilst the message in question was being sent over the line" and thus his client should be acquitted. In fact, the message they hoped to tap did not pass over the line until an hour after their arrest. Judge Murphy, while admitting that the statute might have been more clearly worded, did not think the motion by Fallon's lawyer for the jury to be instructed to acquit him was warrantable and denied that motion. He sentenced Fallon to four years in San Quentin.[22]

In December 1892 California Governor Henry Markham pardoned William Fallon, who had spent some 28 months in custody. Fallon's previous standing in the community was vouched for by, said an account, "many of the best business men" in San Francisco. Frank Jaynes, superintendent at the Western Union Telegraph Company, said it was a case where the executive clemency should be exercised. Markham declared, "It appears that Fallon is a man of good family and has never engaged in any criminal affair; that he unfortunately became a victim of the poolroom craze at San Francisco and lost all of his money." The Governor added that Fallon's 28 months' confinement in the county jail was sufficient punishment. Both this account and the one above said Fallon was sentenced to San Quentin but it appeared he never went there, or was transferred very early on to the county jail. No mention was made of Nagle except to say he had also been sentenced to four years at San Quentin. Presumably he was tried separately as no mention was made of him at Fallon's trial. No published reports seem to have surfaced about Nagle's trial. Nor was any mention made of a pardon for him. It seemed Nagle served out his full sentence at San Quentin.[23]

The Baltimore grand jury indicted Perry (or Terry) W. Wadham in August 1890 for cutting the telegraph lines of Western Union Telegraph Company. He was arrested late that month and charged with tapping the wires. He was found in a room that was well appointed with telegraphy equipment and with which he had established connections with the Arlington racetrack wire. The fraud was said to have been discovered before the scammers had a chance to steal any money. Wadham was released on bail, but fled the area and thus had his bail forfeited. On January 1, 1891, Frank B. Jeffries was arrested in Baltimore and charged with tapping the wires leading to the Arlington racetrack. He was arraigned in police court on January 3, where Baltimore Police Captain Freeburger told the story. Western Union manager Bloxham complained to the police on January 1 that his company's wires had been cut between Baltimore and the Arlington track, about 100 yards east of Oakland Station on the Western Maryland Railroad. In the company of Bloxham, several Western Union linemen, and other detectives, Freeburger traced the wire to a house near the station. Freeburger broke open the door and followed the wire up to a room on the second floor where Jeffries was confronted. He was arrested. In that room he had two telegraph instrument and a battery of 48 cells. On his person he had several letters (unsigned), supposedly from the principal in the scheme, for they gave him instructions and directions on how to proceed, and so forth. Jeffries was held on $1,000 bail. Freeburger was also involved in the Wadham case. Nothing more was reported on Jeffries.[24]

W. K. Wade and Frank Edmunds (telegraph operators), William Johnson (lineman), and Mary Rayson (who was to place the bets), all from New York City, were arrested in Washington, D.C., in September 1890 on the charge of attempting to swindle the bookies by tapping the wires. All intimated they were in the employ of New York City parties. The operators said they were paid $70 a day, while the lineman was receiving $50 and the woman got $50 a day. A house had been rented near the pool rooms. On the morning of September 17 the lineman tapped a wire that proved to be the wrong one. This was discovered at the Western Union office and the wire was repaired. On the morning of September 18 another wire was tapped, also a wrong one. The matter was reported to the police who, after some searching, located the parties and arrested them, and seized several trunks in the room containing a telegraphic outfit, 54 battery cells and a quantity of clothing. That scheme failed because of the bungled work of the lineman. A second account of this scheme gave a quite different version. In this one Johnson received $125 a day and the two operators were to get $25 a day plus a percentage of the profits. Also stated was that the tap was successful, but over a week earlier officials in New York notified the Washington office of Western Union that tapping would take place and a patrol of linemen were detailed to monitor the situation. All four of those arrested were taken to police court on the morning of September 19, declared to be suspicious persons. All were released on promises from each that they would immediately leave the city.[25]

U.S. Marshal Hitchcock left Chicago on the evening of December 17, 1890, for Washington, D.C., with Thomas P. Dudley and Miss Maggie Thompson in custody. They and their confederates in Washington were indicted for defrauding various pool rooms, with the Western Union being the complainant, as it was their wires that had been tapped. An investigation followed the company's complaint and the gang of scammers was tracked down, having made off with an unstated amount of money from the bookies. The pair who got as far as Chicago were arrested there on December 16 and the next day they were brought to court where federal judge Blodgett issued an order that they be taken back to Washington for trial. There was no report on the outcome.[26]

A big swindle took place on January 5, 1892, and the pool rooms of St. Paul and Minneapolis were among those victimized. The racetrack used was the Guttenburg facility in North Bergen, New Jersey. The gang involved in the plot was widely scattered with people placed in St. Paul, Minneapolis, Kansas City, Omaha, St. Louis, and possibly other cities as well. It was estimated that about $50,000 was taken. Three or four men were sent to each of those cities to play the marked horses and divide the profits. On Monday, January 4, four "sporty-looking" men showed up in St. Paul. They said they were from out of town. They were able to work their fraud on two separate races and it was not until the last race was over that the fraud was discovered. Word came from Chicago over the wire that a fraud had taken place and a correct list of winners was sent out. The police were called and within hours the four "sports" were located at a hotel in St. Paul. It was necessary to show that this quartet had guilty knowledge of the fraud and while he lacked evidence pool room owner Frank Shaw bluffed them and managed to get back from them the $7,000 he had lost. The four men then left the city for Chicago. It was also learned that bookies in the San Francisco/Oakland area lost several thousands of dollars; Omaha pool rooms were taken for $6,000 and the Denver and Kansas City bookies each lost $3,000. Five men were arrested in Kansas City while two men in Denver got away. The "luck" those two men in Denver had in their bets caused pool room people there to query New York to verify winners. It was at that point the fraud began to be unearthed.[27]

Western Union quickly got involved in the above-cited swindle and conducted an investigation. Superintendent Humstone said nothing positive had been discovered from that probe but that he believed there had been no tapping of wires but "of course, we cannot ever be certain of that." He speculated on "errors in the transmission of messages" perhaps being the cause. The Western Union officials were convinced their quadruplex telegraphy instruments made it impossible to tap wires without being discovered and all of those officials were then convinced that at least one operator was bribed. A New York City Western Union operator, Frank Boyle, and a Chicago operator, John Graham, had both been suspended during the investigation. A reporter noted that pool room proprietors were notoriously slow to acknowledge they had been swindled and many thought their losses were even greater than stated.[28]

A long piece published in a New York newspaper on March 27, 1892, took a detailed look at the problem. It began by saying an organized gang of wiretappers operated from New York City at the time and it consisted of about 25 dishonest telegraph operators and as many more dishonest linemen, who all tapped and defrauded: "They are not amenable to any law and they know it. Therefore they drift about the country, tapping wires promiscuously, and making a good living without doing much work." Some were said to be based permanently in New York City. The leader of the gang was reported to be a well-known individual who knew all the telegraph operators in the city. One attempted prosecution in New York had recently failed when the effort of Western Union to convict the men was unsuccessful: "No law on the statute books was applicable to their case. Although there was no doubt that the men occupied the room to which the wire ran, the company could not prove that they actually cut the poolroom wire." They tapped the wire and led it into a room they had rented near the pool room, so bets could be quickly placed. The preliminary matter about the races they allowed to pass over the wire unmolested while listening to it and gaining any knowledge they could from it. (Those telegraph wires did not just transmit the odds and results but also items such as weather conditions, starter's delays, false starts, horses getting skittish in the box, and so forth. Those messages were received and interpreted from the Morse code and then read out to the pool room patrons.) When the horses were at the post the tappers began. As all tappers did, they divided the wire into two circuits, one of which ran from the telegraph office to the tappers, the other from the tappers to the pool room. When the news came from the track that the horses were at the post it was sent to all pool rooms on that wire. Nothing more came from the track until the horses were off. During that brief period the bookies were still accepting bets. Horses could be at the post for one minute or for 15 minutes. During that period of uncertainty tappers were able to do their work. When word came from the track that the horses were off the tappers held that message back. As soon as the winner was flashed the confederates of the tappers hustled to the pool room and placed good-sized bets. After waiting a reasonable time for the confederates to get those bets placed the tappers sent out the "They're off" wire to the pool room. A number of minutes later the tappers sent out the same race results that they had been holding back for some 15 minutes or so. Their confederates in the pool room cashed their tickets and quickly left the place.[29]

According to one article, years ago when telegraph companies introduced their service between the racetracks and the pool rooms very little vigilance was exercised over the operators employed by the telegraph firms. Occasionally an operator took advantage of his knowledge. Results arrived at the telegraph office and then were sent on to pool rooms. Such an operator might hide the message on his desk for a few minutes, send out a signal to a confederate in the street, and then send the wire on to the pool rooms. After such practices came to light Western Union watched its employees much more closely. Then tapping became more common and a number of out-of-work operators saw a chance to make money. That worked for a time. Then the telegraph operators introduced a cipher. That caused the tappers to use another plan, "one that would not be operative if the bookmakers used a little more caution." Under the system then in use by Western Union it would be "utterly useless" for tappers to intercept messages as they passed to and from a racetrack and the company's main office in the city because all those messages were in cipher that was known to only two men at the head of the race department at Western Union's headquarters building at 195 Broadway. News of positions of horses was flashed to the city from the track for the information of bookmakers and their customers. No tickets were cashed, however, until confirmation was received. That confirmation came in cipher. The man at the track arranged the cipher message and sent it to his colleague, who compared it with the preliminary report and if it tallied he sent it to the pool rooms. It was said to be a "perfect system" and could not be beaten between the racetrack and the main office of the telegraph company. But it did not prevent the tappers from tapping the pool room wires between the telegraph office and the bookie joint. This account concluded by observing that it was understood that the New York State Legislature was to be asked to pass a law under which a telegraph company could prosecute the wiretappers. "When that is accomplished wire tapping will become a lost art, but not until then, the officials say."[30]

Theodore Diffendorf, 29, for seven years a telegraph operator employed with Western Union, was arraigned at Jefferson Market Police Court on May 18, 1892, charged with the larceny of $23,000 from the bookies at Saratoga, New York, during the races of 1890. At the time Diffendorf, John Harris, and three other men went to the Columbia Hotel at Saratoga and rented rooms on the top floor. They also rented a barn in the loft of which was placed a full operating telegraphy outfit. Then they tapped the wires from the racetrack, rushed their bets to the bookies and reportedly made $23,000. The regular operator on that line found something wrong with the wires, Western Union investigated and found the tapped wires. The telegraphy outfit was still there but the men had fled. Harris was arrested on May 9 and Diffendorf a few days later. No outcome was reported.[31]

Robert Smith (formerly a Western Union operator), John Smith (his brother), and Charles Dougherty (a lineman) were in police court in Cincinnati on October 6, 1892, charged with tapping the wires of the Western Union Telegraph Company. However, the prosecuting attorney was indisposed to prosecute. Privately he said the pool rooms were lawbreakers and the city was trying to suppress them and he looked upon wiretappers as efficient assistants in breaking up the illegal business of the bookies. In the courtroom, though, he based his objection to prosecution on the ground of defects in the statute under which the arrest was made. Western Union was described as wanting to "push the prosecution." The case was continued for a few days. That very afternoon the Cincinnati chief of police began a war on the pool rooms. Police were sent out with instructions to arrest as many bettors as they could catch in the act. Several arrests were reported to have been made on that first day of the campaign. It appeared that no prosecution of this case was undertaken.[32]

A piece published in a Washington, D.C., newspaper in November 1892 looked at the world of the telegraph operators who were involved in racing. According to this article, the most fascinating feature of a telegrapher's work was in the transmitting and receiving of racing reports. Bookies were said to generally pay operators big wages, $5 a day for four hours a day with the fair certainty of getting much of that money back by the end of the day—due to the temptation of placing bets while working in a bookie joint. Telegraphers who handled racing reports were said generally to be young men who were experts at the key and sounder in telegraphy. Old hands generally declined the positions. (The article did not say why that was so.) Supposedly the surroundings of a bookie joint had a demoralizing effect upon the moral character of the operators "and the good pay is poor recompense." After they had worked the races for a season or two they were fully acquainted with the workings of the different race circuits and with several of the codes or ciphers in use. "The next season, should they not be able to secure their former position or one similar, they are generally open for negotiations from outside betting men, and a scheme for tapping the wires is generally the result."[33]

This article went on to declare that if it had not been for the vigorous action of the Western Union Telegraph Company in prosecuting the tappers that were captured many more of those schemers would have emerged. The telegraph company derived "a princely revenue" from the pool rooms and had to "protect its customers, as they know the law will not recognize the latter, as they, too, are battling against the statutes." The journalist who wrote this piece dropped into the office of one of the big telegraph companies at midnight one day and had a talk in the corner with several operators about the issue of wiretapping. One operator told a story about something from about 10 years earlier. Operators, it was said, were then paid $10 for four hours of work. One day a message came over the wire in cipher. All operators could read it even though only a select few supposedly could. That message outlined how the third race at Jerome Park was to be won by drugging the favorite, as well as engineering other underhanded work, and named an obscure outsider as the winner. Odds were about 25 to 1 against that horse. It was then near the end of the racing season and most operators had money put away. The story teller said he put $500 on the horse (with his boss) and many other operators also bet heavily on the long shot. However, the favorite won and the long shot finished fifth. One of the bookies involved found the real story too good to hide and told his friends how they had suckered the operators out of much of their saved-up pay by planting that false message about drugging. But then, sneered the bookie, the operators should not have betrayed the confidence placed in them. Another operator told a story about tapping. He, the operator, was at his boarding house in New Orleans, which also housed several racing men and gamblers. One day at the dinner table, the operator outlined a scheme of his as to how a wiretapping could be done successfully. It was to find a wire from the track to the bookies that made its way through a large tree that obscured the tapping site. This operator was later approached by one of the racing men at the boarding house and urged to put the scheme into play. But the operator declined, saying he wanted to stay in the telegraph business as his livelihood. In the end the racing man got another operator and was successful with the plan, scamming several thousands of dollars from the pool rooms.[34]

A comment on the pool room business appeared in a San Francisco newspaper at the beginning of 1893. It declared the pool room was a money machine for its owner but that it would not be long before the pool rooms were closed up. Very little capital was needed to start one up, just a "straw bond" of $10,000 deposited with the County Clerk and nothing else. No character references were needed. A straw bond carried a fictitious name for the guarantor or the name of a person who was unable to pay the guaranteed sum. In other words, it was a wholly bogus instrument.[35]

Just one day later, an article in a New York City paper stated that a bill was being readied to be introduced into the New York State Legislature to abolish certain types of betting in bookie joints and to confine all racetrack wagers to the parimutuel system. Said the story, "In the midst of all the campaigns against wickedness of the island the poolrooms thrive amazingly and escape the attacks of the church as well as the raids of the constabulary. They seem exempt from all good influences." The article added, "The utter defiance of public decency is an outrage on the morality of the metropolis."[36]

Still in January 1893, at the end of the month in New York City, pool room proprietors had reportedly been complaining to Western Union for a period of several months that expert wiretappers had been stealing their returns from racing and robbing them by betting on the stolen information. Wilkinson's detective agency was hired by Western to investigate and they set to work. Around January 23 New York Police Department Inspector McLaughlin ran into one of those private eyes. They met again five days later and then five prisoners were arraigned in the Tombs Police Court on January 29. Several days earlier a man entered the Springer House hotel in New York and went up to an office on the second floor. He met Mrs. Hillen, wife of the owner, and told her he was a Western Union lineman and wanted to go up to the roof to repair some wires. She allowed him to do so. On the same day a young man rented a room on the top floor of that hotel, looking out on a certain street, and sent up a small trunk to his room. At the same time Western Union investigators were watching these men. On Saturday, January 28 those investigators concluded the men were tapping the wires leading into Peter Downey's pool room. They sent word to McLaughlin and he detailed some of his police officers to arrest the men. Those arrested gave their names as John Ward, 24, Thomas B. Russell, 21, Jacob Maguire, 29, Charles Martin, 28, and Richard Smyth. A policeman at the station recognized all but Martin as former Western Union linemen. In that rented room at the top floor of the hotel were a trunk filled with batteries and extra telegraph receiving instruments. A wire had been tapped on the roof and fed down to that hotel room through the window. It had then been connected with a receiving instrument concealed in a bureau drawer. Also in the room were a number of large paper squares with numbers one through 10 printed on them (one to a sheet). After the man in the room took down the message with the name of the winning horse he displayed that horse's number from his window to confederates in the street. During the period of delay before the man in the hotel room sent on the message those confederates had time to go across the street to the bookie joint to place their bets. When Inspector McLaughlin saw Martin he recognized him as an old offender whom he had arrested twice before for wiretapping, in November 1891 and July 1892. He managed to get off from the charges both times. Said McLaughlin, "It is very hard to convict these fellows and I have had men working all day to get additional evidence in this case." Those arrested men had been caught before they started to work and thus they had made no money from the Downey pool room.[37]

An interview was conducted in Chicago in March 1893 with a man named O. M. Stone, said to be a "master of all things electrical" and a man who had caused the Chicago Board of Trade more bother than all the other alleged electricians put together. He had reportedly hatched so many schemes to circumvent the powers that be that he had been called "the king of the wire tappers." A reporter from the *Chicago Herald* asked Stone whether he thought wiretapping could be accomplished. He thought it could but it would require, in Chicago at least, a knowledge of electricity that not one in a hundred possessed. He said he thought about 100 different parties had approached him in the previous three years asking him for information on the subject of wiretapping. "And it has invariably been in regard to horse racing that they wanted enlightenment." He said he always advised operators with expertise in telegraphy not to go in for it. He thought a large proportion of those who tapped—that is, operators who thought they knew more than they did—made "miserable failures." Stone was asked how he thought a wire could be successfully tapped. He replied that a single wire was the easiest to tap, and the most common type to encounter. A duplex wire was the next most difficult and then came the quadruplex wire, which meant the tappers had to have four men at each end of a wire, all working at the same time. But Stone assured the newsman that it could be done.[38]

In the very early morning hours of May 2, 1893, Fred Barton (night clerk at the Victoria Hotel), William Willard (an employee of the Citizens' Electric Light Company), Ed Cameron (a Western Union lineman), B. Richardson (a telegraph operator), and W. W. Snyder were arrested and placed in jail in Louisville, Kentucky, charged with "interfering with and obstructing the service of a telegraph company," which was a felony, according to the laws of Kentucky. They were arrested by the police after several days of investigation. Some days earlier Chief of Detectives Owen received information that the local pool rooms were being made the victims of a gang of wiretappers and were losing large amounts of money. It was learned the wires were tapped near a train station, the line over which the reports of the races at Nashville and Lexington were sent.[39] At the end of June those five men stood trial in Louisville Circuit Court. All were convicted except for Cameron who had his case dismissed. Richardson

An 1893 sketch of the hotel room used by the tappers. The tapped line came from the roof and down through the window into the hotel room where it was hidden in the dresser drawer. When the tapper held back the race results he dropped a numbered card out of the window. A confederate picked it up, raced across the street to the bookie joint and bet on the horse corresponding to the card's number.

was fined $250 while Barton, Willard, and Snyder were each fined $150. All the fines were paid on the spot and the men were all discharged.[40]

On March 29, 1894, the story "leaked out" that Western Union wires running between pool rooms in Kansas City, Kansas, and Kansas City, Missouri, were tapped some time on Wednesday, March 28, and fraudulent bets made on races running at East St. Louis, Illinois. The cities' bookie joints were taken for an amount estimated to be from $4,000 to $5,000, with the biggest losers the pool rooms of Jack Dugan and those of the Maltbys. The perpetrators were never caught.[41]

Captain Schmittberger, along with three detectives on the New York Police Department force, broke up a wiretapping scheme on the afternoon of April 14, 1894, in New York City. Western Union had noticed there was something wrong with two of their wires on the West Side on Friday, April 13, and linemen were sent out to discover the problem. The wires affected were those over which reports from the East St. Louis and Memphis racetracks were received. It was suspected, therefore, that somebody was engaged in tapping. By chance, Schmittberger and a couple of his detectives noticed a pair of suspicious characters near a jewelry store. Thinking they might be planning a jewel robbery, they watched the pair. After a couple of hours the pair moved off. Since there was no robbery the police stopped watching but did report at the station that the men were evidently watching the roofs of houses. On the morning of April 14 Schmittberger heard from Western Union that there was a leak in the line in the area near that same jewelry store. The captain concluded those suspicious men were watchers for the tappers. A policeman was on hand when the same men appeared at the same place at noon on the 14th. At 2:00 p.m. the men entered the basement of a house, a former pool room kept by a man named Beaman. More policemen were called in. That basement was broken into. A wiretap operation was found to be in place in the back room, complete with desks and telegraph instruments. The men involved were all arrested, four in total: John McNally, 21, John Sweeney, 27, Frank Baird, and Joe Cotton, 55. In the room the police found two relays with keys and sounders, an extra sounder and two valises. One of those contained 21 battery cells while the other contained a portable switchboard with 12 cut-out plugs. Office supplies and equipment were also found, along with racing cards with the entry lists for East St. Louis and Memphis races and a cipher code. When Schmittberger cut those wires at the tapping room he inadvertently stopped all the business on the wires on the West Side circuit. He soon heard about his error, though, as a Western Union official was at the police station soon after the raid on the basement. Western Union linemen were sent out and the lines were quickly repaired. Charges against Cotton were dismissed but the other three were held for trial. No outcome was reported.[42]

A report in April 1894 exclaimed, "The extraordinary activity of the Western Union Telegraph Company in the matter of bringing the wire tappers to justice has often excited remark. The company employs numerous detectives, who are constantly watching all the lines where the race wires are strung, and a general system is in operation in the racing department of the company which involves the expenditure of a very large amount of money yearly." According to this account, wiretapping was lucrative but a practice that was never successful for any great length of time, due to the vigilance of the telegraph companies. One theory was that all the effort by those firms to stop tappers was done in the interests of morality, "but as a matter of fact it is a part of the contract that the telegraph company makes with the bookmakers." Western Union paid the racetracks for the right to report results; it was that information that was then sold to pool rooms for large sums, in New York City said to be as high as $100 a day but governed somewhat by the condition of police activity. In other places the price was lower. Bookies paid high prices for the race information but also demanded protection. The information was only valuable when it was exclusive and trustworthy, "Hence the tireless vigilance of the telegraph company whenever there is danger that the wires will be interfered with."[43]

A report in August 1894 from Columbus, Ohio, about wiretapping reached the Western Union office in New York City through a reporter. Those Western Union officials had been watching closely for schemes of that sort and were not surprised to hear of a scam. Those named in the report as involved were operators J. G. McCloskey and J. H. Mittleberger, who were both well-known at Western Union headquarters in New York City. At least one of them had been under suspicion before. There was also a case of wiretapping in New York State on Friday, July 13, 1894, involving the wires running between New York and Albany. It happened that racing news was carried by wire to Canadian points by way of Albany. The operator there discovered reports were not running regularly. He could hear reports from Chicago going to New York City but in the transfer to Canadian points the reports appeared to be unusually late. After calling the attention of the New York office to that fact and after an investigation was conducted, it was determined the wire had been cut. Immediately a new wire was supplied between Albany and New York City for the Canadian points and a change was made from that city in the way winners were announced over the wire for the benefit of the Canadian pool rooms. Then the tappers were informed that they had been detected. A message sent to them by the manager from New York City said, "Boys, I am sorry, but we may as well discontinue operations, for we have all your friends' money and they are broke." An "ugly message" was sent back by the tappers. No detail was provided about the above-mentioned Columbus case except an August 15 note that McCloskey and the other New

York City men arrested with him (Kendall and Martin) and charged with attempted wiretapping had all been released by Judge Hagerty in Columbus because he found the Ohio state law defective inasmuch as it did not constitute an offense to tap a wire. Also, there had to be a message taken from the wire.[44]

Another wiretapping scheme was unearthed by Western Union officials in September 1894. It was located in Washington, D.C., and involved racetracks in Virginia. The company's wires that were involved ran over the Long Bridge over the Potomac River. For several days in that month Western Union officials kept themselves informed about the doings of a couple of suspicious men who were hanging around the bridge and its immediate area. Part of the information Western Union was working with involved the fact the men had been seen out in a boat on the nights of September 11 and 12. On the afternoon of the 13th a sudden unusual clicking was heard at Western Union's central office in Washington. Two of the firm's officials were immediately sent to the bridge. At first they saw nothing but then they noticed the two men in a sailboat about 100 yards from the bridge. It looked like the men were fishing. Finally one of the Western Union men found the tap wire running down from the pole and down the bridge structure. Pulling on that wire showed it led to a cable in the water that was attached to the boat. By that time the police had been called and had arrived on the scene. However, the act of pulling on the cable had alerted the men in the boat and, although they were pursued by the police in another boat, the two tappers escaped after stealing a pair of handy bicycles. One of the Western Union officials, J. W. Collins, observed, "It would have been necessary for the tappers to have first familiarized themselves with the New York operator's telegraphing, for it is a fact that every operator has a certain style, which is easily apparent to a receiver." Collins thought the tappers had confederates stationed in pool rooms who could easily be tipped off to the winner of a race by means of a certain dash or series of dashes, equivalent to a cipher, which they could understand without the regular pool room operator catching on.[45]

With respect to the Washington scam, it was also reported that Western Union had resorted to a more thorough system to prevent any advantage gained by wiretappers. Heretofore the sender of a race result simply telegraphed the winner and other details by the regular Morse code of the transmission. But now a cipher had been added with the result that the receiver of a message replied with either the first or last two or three letters of the cipher and the remainder was supplied by the sender. Because of this cipher Western Union felt that tappers would be unable to supply the missing letters and any delay in answering would be sufficient grounds for suspecting something was wrong. It this case it had apparently not worked.[46]

Within a week the culprits in the Washington case were under arrest. Fred J. Owens and Walter Geddis were the two men in the boat. A third man by the name of Palmer B. Babcock was also under arrest for his involvement in the scam. Babcock was a telegraph operator. A fourth man who was also involved the scam was said to have left town. The three in custody were said to be familiar faces at the area racetracks. All were charged with conspiracy but, wrote a newsman, "there is little probability of their being convicted, because no law adequately covers the case. Geddis and Owens may be convicted for stealing the bicycles on which they escaped, however."[47]

A further Washington newspaper article elaborated on the dilemma that confronted police officials when Geddis and Owens were arraigned on a charge of grand larceny. "It is not thought probable that the prisoners can be tried for this offense, although they are said to have practically admitted their guilt," said the article. "No law has been found which will cover the case, although the lawyers in the district attorney's office have made an extensive search of authorities and possible precedents." Also noted was that several similar cases had come up in the courts but "it was always found impossible to punish the swindlers." Owens was the one who hired the sailboat while Geddis acknowledged buying the cable that was used in making the illicit connections.[48]

Two cases of interest in racing circles came up before Judge Cole on January 5, 1895, in Washington, D.C. Palmer P. Babcock and F. J. Owens were arraigned for wiretapping. This, it was alleged, constituted a conspiracy to defraud and injure the Western Union Telegraph Company, whose wires were tapped. The two men pled not guilty, with leave to file a demurrer, an assertion by a defendant that though the facts alleged by the plaintiff in the complaint may be true, they do not entitle the plaintiff to prevail in the case. It was understood they would contend that as bookmaking was an offense against the law, Western Union was engaged in an unlawful business in transmitting information to bookmakers and was, therefore, unable to assert that the wiretapping injured the company's legal business. A second case, also with a demurrer, was to argue that bookmaking was not unlawful in the district, contrary to another court decision that bookmaking was unlawful under the laws of the District of Columbia. The two cases were unrelated.[49]

More than one year later, on February 16, 1895, Judge Cole decided that it was unlawful to tap telegraph wires, whether they were being used for a lawful purpose or not. Attorney Thomas Taylor, acting for Babcock and Owens, had tried to quash the indictment, arguing that the wires were not used for any purpose whatever except to convey information for the conduct of bookmaking businesses, with racing as its basis. As bookmaking was illegal in the District, the wiretappers did not interfere with a lawful business. Judge Cole overruled that argument and also added that "if the wire tapping had been by [police] officers for the purpose of breaking up the unlawful business it might have been a different matter." No more was reported on this case.[50]

A gang of wiretappers operating in New York City's Tenderloin district for a week or more were run down on November 20, 1894, but not until they had taken in thousands of dollars. The arrests also disclosed that no fewer than six bookie joints were running in the Tenderloin, a fact that was said to be a surprise to the New York Police Department officers in that precinct, especially to Captain Schmittberger. Arrested were Augustus Araushaar and Jacob Bessinger. They operated out of a room Bessinger rented in a rooming-house. He had so many callers that the landlady got suspicious and called the police. They investigated and discovered a wiretapping operation. After collecting evidence the police launched a raid and arrested the men. Among the evidence found was a wire coming in the back window of the room, and a set of telegraph instruments. Bessinger was operating the key when the police conducted their raid. Nothing more was reported.[51]

Officials of the Pacific Coast Jockey Club in San Francisco charged in January 1896 that the Western Union wire from the city's Ingleside racetrack, which was leased by the Jockey Club, had been systematically tapped and the race results given to all the pool rooms in town. Discharged telegraph operators were suspected of being the tappers. In its fight against tappers the racetrack management had closed the telegraph office at the track with the result that no offices on the coast would be able to get the racing results from the track.[52]

Texas pool rooms were reported to have been hard-hit by a wiretapping scheme in March 1896. Houston, Galveston, and San Antonio bookie joints suffered losses of over $2,000. It was a scam that extended outside of Texas as Chicago pool rooms lost between $7,000 and $8,000. Thousands of dollars were taken in Louisville, Kentucky. Nothing had happened in San Francisco because there was reportedly not a single bookie joint then in operation. Denver bookies sustained losses of about $1,600. That scam was based on the last race at a New Orleans racetrack. Officials of the Western Union company in New Orleans were said to be "reticent" concerning the false wires sent out by the tappers, "but from the best obtainable information the trick was turned by an operator in the telegraph office at the race track."⁵³

Another report of this scam called it one of the biggest, noting the conspirators had agents in many cities where gambling on racing was carried on and that large bets were placed. Plans were complex, with money being distributed by telegraph from New Orleans and Chicago to those agents 24 hours in advance of the targeted race. When that last race was finished at New Orleans the news was flashed all over the country that Royal Nettie had won. Soon after bets had been paid, suspicions of some bookies were aroused, but it was then more than an hour after the race was over that they learned a horse called Plug had really won. In this account the Chicago losses were estimated at from $15,000 to $18,000. Owing to the reluctance of pool room managers to discuss the subject it was felt impossible to give an exact total of the losses. Reportedly $5,000 was sent to Louisville from Chicago the day before, to play on the race, and $1,000 was telegraphed to Chicago from New Orleans. At New Orleans more speculation continued to swirl around the idea that an operator at the race track was responsible for changing the race result. Said a reporter, "It is stated by telegraph experts that the wire could not be tapped, as the wire to the racetrack is what is known as a duplex wire and an intermediate set of instruments could not be put in without attracting instant attention." Western Union officials continued to refuse to make a statement. In Covington, Kentucky, there were said to be no pool rooms known to the law, but several institutions of that nature doing a regular business undercover were hit in the scam to some extent. That information came to reporters from outside sources and not from management.⁵⁴

One day later an article describing the above scam placed the loss at from $250,000 to $500,000. The total could be increased significantly after the estimated 280 pool rooms then running in New York City gave their accounting. Speculation was that the fraud was planned in New York City and carried out with New York money. The total loss in Chicago was put at $27,000 with the pool room owned by James O'Leary suffering the single largest loss, at $10,000. Of that amount, the O'Leary establishment paid $7,000 in a lump sum to a stranger who had placed a $1,000 bet with him.⁵⁵

By then Superintendent West and other officials of the Western Union Telegraph Company were busy trying to figure out what had happened. It was said that no leak was discovered between the telegraph office and the racetrack and that West was convinced the false report was sent out by his own operator at the track, although whether by accident or design West was not prepared to say. Three officials of the Jockey Club called on

West and in their presence he laid the blame on James Conway, one of the firm's operators at the racetrack. Conway declared he had sent off a message with the correct winner of the race in question and if the result was changed that alteration took place after it left his hands. Conway insisted the wire must have been tapped. The wires were, Conway added, "fluttering and in trouble" all day. One of the two wires leading away from the track had gone wrong in the morning and Conway said he was not able to use it all day long. Conway had been with Western Union for four years and declared, "The idea that I was in with any crooked play is ridiculous…"[56]

William H. McNutt, described by a reporter as "the most notorious wire-tapper in the United States," was arrested in Chicago on March 19, 1896, by the Chicago police on suspicion of being implicated in the above-cited fraud. This account put the total of the loss at $150,000. McNutt was locked up at the Central Station but not booked. The prisoner declared he had nothing to do with the fraud, but knew all about it. It was speculated by the reporter that the police did not think McNutt was a professional wiretapper but a man who raised the money to hire people to do the job.[57]

Two weeks later there had been no arrests. A lengthy article looked at the issue of wiretapping from a larger perspective. It said the thieves who resorted to tapping were always expert telegraphers. The fact that there had not been more than eight or ten such dishonest operators out of the thousands in the United States was said to have been a mark of the overall honesty of those in that profession. All official racetrack information used by bookmakers was supplied to them by the Western Union Telegraph Company. When a man wanted to open a bookie joint, whether it was in New York City, Hoboken, Chicago, Pittsburgh, or Long Island City, he applied to the racing bureau of Western Union. That bureau ran wires to his establishment, supplied him with operators and furnished him daily with the entries, jockeys, condition of the track, odds, and other information, from whichever tracks he was interested in. The moment the horses were at the post the telegraphic instrument in the bookie's place notified him. In the same way he learned of the progress of the race and, eventually, the name of the winner. Upon receipt of the latter the bookie paid off the winning ticket holders. All that information, and other items from the track, was telegraphed directly from the track to the Western Union racing bureau in New York City. From there it was reported to all the bookmakers who got the bureau's service. The aim of the tapper was either to cut in on the line between the track and the racing bureau or to cut in between the bureau and the bookmaker to whom the news was sent. If it was the former the tapper cut off the racetrack operator and, imitating the cut-off man's style of sending, gave the bureau the name of the horse he and his confederate had agreed to back. If the latter method was used the tapper imitated the style of the sending operator at the Western Union racing bureau and sent out the false winner. The account went on to say that the former method had not been successfully used in many years—one could even say it had gone out of fashion as the risk of detection was too great.[58]

This reporter went on to declare that in this particular fraud "there was no tapping." The operator at New Orleans simply sent out the name of the fake winner, Royal Nettie, instead of the real winner, Plug. By this time Western Union had suspended Conway and Maguire (another of its operators at the New Orleans track) from duty. It was also noted that most of the "wise men" who made hand books in New York City (that is, who were not connected with the telegraph system) had a custom of not paying bets until the day after the race and thus suffered no losses from tapping scams. Reportedly, in cities around the country the bettors who had wagered on Royal Nettie had been placing small bets and losing with regularity for a week or more before that fateful day. Thus, they were much less likely to raise suspicion. And, wrote the newsman; "To send confederates out in this way 'piking' along for days or weeks to prevent suspicion is a regular part of the wire-tappers' plan of operations." Old-time telegraphers knew that it used to be impossible to cut in on a duplex line—one in which two currents were constantly traveling in opposite directions—but that now some could tap a duplex line successfully. Tapping was said to be getting more difficult in New York City because the wires ran in cables underground or along the elevated railroad structure. According to this report, the great difficulty in wiretapping was to imitate the style of the operator who sent out the legitimate racing reports. No two operators sent messages alike: "There is as much individuality in sending as there is in singing, talking, or playing the violin. It is far easier to forge a man's handwriting than it is to imitate perfectly his style of sending Morse." Also, those legitimate racetrack telegraph operators were extremely fast, so the tappers had to be equally speedy as well as being good mimics.[59]

Charles Moran, 23, and H. M. Summerfield, 26, who had been posing as expert electricians employed by Western Union of New York City, were arrested in Denver on June 2, 1897 on charges of wiretapping and swindling pool rooms in Denver and other cities. Their rooms in an office building were raided and complete and expensive wiretapping equipment was found. The men did not deny their business was robbing pool rooms or that they operated in England, Germany, Canada, and all over the

United States, but they claimed they had done nothing of that kind in Denver. But it was said that pool rooms in Denver had been hit that day with losses of $3,300 and that those two men were the perpetrators. Western Union was reported to be making every effort to secure evidence against them from other cities in which they were supposed to have worked.[60]

Two days later, on June 4, Moran and Summerfield were released from custody for lack of evidence. They were free in talking to reporters about their supposed prowess in other cities. Summerfield said he won a telegraphy contest in 1891 as a fast sender. He declared, "Five years ago Moran and I went into the business of tapping the wires of pool rooms and race tracks. We saw how easy it was to make money in that way. The pool rooms, you know, have no legal existence and therefore they cannot punish us." He continued by stating that in New York the pair used as a blind an x-ray apparatus factory under the name of LeRoy and Company. "Sometimes we got local parties to place our bets on a 10 percent commission to avoid suspicion." According to him, the pair had robbed pool rooms in Germany, England, and the United States in the previous five years to the tune of about $500,000. In St. Louis, he said, they had done four bookie joints out of $7,000 two months earlier. Summerfield remarked that the pair had been arrested in New York on wiretapping charges in October 1890 and in July 1891.[61]

Moran and Summerfield were known to both the New York Police Department and to pool room keepers in New York City. They were believed to be the men who had made $600 about two months earlier by tapping the wires into Little Monte Carlo at Weehawken, New Jersey. At that time they delayed the returns of a race and substituted a different horse as the winner, on which their confederates had bet heavily. Those confederates cashed their tickets and when the fraud was detected one hour later they had all left. Summerfield was said to be the chief planner and, said a newsman, "It is said of him that he is able to imitate the manner of sending of other telegraphers so well as to deceive men who have been accustomed to one another's ways for years." As well, a gang, presumably managed by Summerfield had been winning small amounts from pool rooms in the neighborhood of Columbus Avenue and 59th Street in New York City for a year or so. So persistent and successful were they that the pool room keepers would refuse to pay bets made at long odds until the result received over the wire had been confirmed by a telephone message to some point outside the wiretappers' range. As the men had boasted in Denver, the last thing the pool room owners would do was to appeal to the police in regard to their illicit operations.[62]

According to the New York Police Department the last record of

Summerfield in their city was that of his arrest on August 12, 1891, under the name of Lawrence Somers, on a fraud charge that was not wiretapping. Police had no record of his arrest in 1890, although he could have been booked under a fictitious name, something that commonly occurred before fingerprinting and rogues' gallery photos made it easier for police agencies to attach various names to a particular criminal. The police did admit that Summerfield had made a great deal of money in the wiretapping business but laughed at the idea of $500,000 in five years. Not taking this pair seriously as well were the Western Union people. The story of operating an x-ray studio as a blind, those officials said, showed the pair to be liars, because there had been no pool rooms worth fleecing in that neighborhood since the invention of x-rays.[63]

Taken into custody by Louisville, Kentucky, police detectives in November 1897 were three men thought to be planning a raid on the pool rooms of that city by tapping a wire at Lexington. The leader of the trio was Charles Moran, onetime owner of a string of race horses and reputable New York turf man. C. M. Butterfield from New York, an "expert telegrapher" at Western Union, was the second man in custody; Luke Burke was the third. All three were said to have been in custody in Denver at the beginning of June on wiretapping charges but they had posted bond and then skipped town before their trials, thus, of course, forfeiting those bonds. On Saturday, November 20, the men, all being elegantly dressed, made an appearance at the Turf Exchange, made a few bets and mingled with the crowd. It was said they might have succeeded except somebody in the crowd recognized them and tipped off the police, who arrived at the pool room and arrested them. They all denied they were the wiretappers from Denver but admitted it when they were confronted with their Denver mug shots. All were remanded to jail but afterward were given bail.[64]

Police in New York City seized $700 worth of wiretapping equipment in a room on West 44th Street on December 23, 1900. They narrowly missed the gang whom they thought were the same men who had also newly been arrested by the Chicago, Omaha, and St. Louis police. Evidence was found that the gang planned to tap the Chicago to New Orleans wire. Found in the room were seven telegraph instruments, a sounder, eight dry batteries, three wiretapping machines such as were used by telegraphers in checking wires for breaks, and a quantity of wire. The wire in the room was connected and ready to go but it was not then connected to the Chicago to New Orleans wire. An unaddressed letter police found in the room read, "Don't fail to wire Great Northern, Chicago, second signal. If horse is scratched, better substitute another, as it might make some difference in paying, as you know. A false winner is not so easy. Be careful with your ciphers, and take the last race of the day. I must hurry, in order to catch a train on the Lake Shore. Good luck and be careful." Also found in the room were some newspaper clippings from western papers about wiretapping gangs having been captured in various cities.[65]

When the Covington and Newport, Kentucky, pool rooms were hit on February 7, 1902, for more than $20,000 in losses, there were many guesses as to how it was done. It was then stated the next day that wires had been tapped at some point north of Dayton, Ohio, as other pool rooms were hit south of Cincinnati but none north of Dayton. Most of the money from the pool rooms was placed by visitors from Dayton. Pool room men claimed that their reports were delayed 15 minutes, during which time each of four pool rooms was hit for over $4,000 on the last race of the day run at New Orleans. Western Union and the pool room men were said to have initiated an investigation. In this instance, as was often the case, the police were not called in. If and when Western Union and the bookies thought they had solved the mystery then they would call in the police and turn over whatever evidence they had.[66]

When he exposed a wiretapping operation in Butte, Montana, in February 1902, Charles W. Clark, son of U.S. Senator William A. Clark, saved the pool room operators in that city a sum estimated at from $25,000 to $30,000 on the races run at Oakland, California. Parties involved in that alleged fraud had escaped and were thought to be in Utah. The results of those races were delayed by wire and the information phoned to Butte about 15 minutes in advance of their receipt of the telegraphic information. Clark, it was reported, received a tip from a New York City man who was then in Butte. To ascertain the truth of the tip and the extent of the wiretapping, Clark placed several thousand dollars on the result of one race. Several men in the conspiracy also put up large sums; they stood to break the pool rooms but for the fact that Clark, after satisfying himself that the fraud was being carried out, successfully exposed the plot and pulled out his money. "I like to make money easy, but I am not that kind of sport," Clark said. He had stood to win about $10,000. No bets were paid on the race and the money placed by the alleged wiretappers was still held by the pool room men. On the day the fraud was exposed, the wires between Ogden and Butte went down for about 15 minutes and it was believed it was then the conspirators got in their work. By causing the wires to go down for a short time just prior to the time of the telegraphic arrival of the race results, the reports could be received in Salt Lake City and telephoned to Butte several minutes before the wires could be put into working order again.[67]

A gang of wiretappers, thought to be six in number, tapped a wire leading from the Western Union office to Black & Fitzgerald's pool room in Los Angeles on September 13, 1902, and by withholding the race results fleeced the pool room out of an amount variously estimated at from $1,500 to $4,000. The amount of money suddenly bet on the winning horse in a specific race and the delay in receiving the wire with the results being about 30 minutes, the suspicions of the pool room owner were raised and betting on that race was closed earlier than was the custom. However, said a report, "When the result finally came in there was no suspicion that the delay had been caused by wire-tapping, and most of the bets were paid." Some of the larger bets, though, were not presented for payment. Pool room operators in the area protested to Western Union manager Beardsley in Los Angeles about the long delay in receiving the returns on that particular race. When he investigated and learned that there had been no delay in sending through the Western Union he had the line examined and found where it had been cut. Police were called in and went on to state that six men were in on the fraud and that three of them were known.[68]

A few days later, Charles L. Mattfeldt was arrested at Long Beach on suspicion of being involved in the Los Angeles scam. According to this account, the take was $2,800. Since the fraud had taken place Western Union had its own investigators along with private detectives at work on the case. It was said that Mattfeldt had been in the room to which the tapping wire was traced. No more was reported on this case.[69]

Pool rooms in New York City and throughout the eastern United States were reported to be having trouble at the start of 1903, and it was not just through the attitudes of the police, who were always trying to shut them down. A "select company of people" had been winning large sums of money with a regularity that was "astonishing" to pool room operators. A certain number of players were remarkably successful in picking the winners of the races held in California. At first, only small bets were placed, but as time passed the bets grew larger. Also, those customers picked only winners when they bet on the California races and made their bets only a minute or two before the wire announced that the horses were off and all bets were closed. The amount lost by the bookies was unknown but thought to be in the thousands of dollars. In this case it was said that the trip across the continent required seven minutes for a telegraph message. "Certain people found a way of getting this information from California in about three minutes. This gave them a leeway of about four minutes to beat the poolrooms. In most of the rooms bets are received right up to the telegraphic flash that the horses are off."[70]

With the arrest of J. H. Poindexter in Chicago on November 3, 1903, the police claimed to have uncovered what a reporter called "the greatest wire-tapping conspiracy ever encountered in the West." It contemplated swindling the pool rooms out of enormous sums. When the police launched a raid a week earlier it had netted them $10,000 worth of telegraphic equipment and knowledge of a scheme designed to swindle Dallas, New Orleans, Galveston, Houston, Hot Springs, and numerous smaller cities. The correspondence of agents in Cleveland, St. Louis, St. Paul, and Cincinnati was also found. Police declared that with the arrest of Poindexter they had a complete case.[71]

Another one of those arrested in Chicago was O. M. Stone. Letters addressed to Stone, which were found when he was arrested, showed that many successful tapping scams had already been executed in which sums varying from $1,000 to $25,000 had been secured. Telegraphic outfits had been bought and had been shipped to the different cities where the wires were to be tapped, including, besides those cities listed above, San Antonio and Waco. One of the letters was posted from Hot Springs and was from Poindexter to Stone. Dated January 17, it read, "I have entire access to Western Union office here. I am in the electric railroad scheme with Ryan, the manager [to float the bonds]. Now, here's our chance. The poolroom cable comes out of the office in rear of building. I can rent offices so we can reach out of window and handle it. We can cut 'em off for one or two minutes and there will never be a tumble. The money is being bet like wildfire here."[72]

So well known was the wiretapping scam that was worked on the races that a fiction book was published on the subject in 1906, *The Wire Tappers* by Arthur Stringer, published by Little, Brown and Company, Boston. According to one reviewer, the "scenes are almost entirely in the underworld of graft and crooks." Lead characters in the book were an expert in telegraphy (an ex-convict), an inventor, and a skilled electrician. They comprised the gang engaged in wiretapping and beating the bookie joints. "There is a beautiful English girl in the gang who is much stuck with Durkin (the ex-convict) and he with her." That was very much unlike real life, as in all the articles about real wiretappers women were always absent. This book was reviewed in many different publications. Even though the characters were all criminal low-lifes and the ex-convict and the girl were madly in love with each other, a reviewer did note, "through it all the girl preserves her chastity."[73]

A wiretapping scam that netted $200,000 and took anywhere from $25,000 to $50,000 of that total out of New York City pool rooms was worked on the afternoon of July 5, 1906, in connection with the telegraph-

4. Wiretapping the Bookies

ing of the results of the second race at Windsor, Ontario. The story, as passed among the gamblers (apparently the police were not called in), was that wires were tapped at Utica. Reports came from Rochester, Buffalo, Fort Erie, Albany and Schenectady that pool rooms in those cities had been hard-hit. Louisville, Kentucky, bookie joints lost $2,000. Messages confirming the fake race result came three minutes after receipt of the message announcing the winner. This aroused suspicion, as the confirmation of a race winner was usually not received until after the posting of the betting of the next race—a much longer period than three minutes. In those pool rooms where suspicions were raised quickly enough bettors were told they would have to wait until the next day before their bets were paid. Reportedly, New York City pool rooms

A pair of 1906 ads for the novel *The Wire Tappers*. In the book a woman was featured prominently as one of the gang of wiretappers. In the real-life world of tappers no women were in those gangs.

The Wire Tappers
By ARTHUR STRINGER.

No more exciting story has been published for a twelvemonth. The theme is fresh, the author's method is virile, the characters, though criminals, are altogether human, the action of the tale is rapid and each chapter has a climax that is worth while. If you enjoy a series of startling adventures, all with an element of possibility, read of Frances Candler and Jim Durkin and their struggle with MacNutt and with the world.—*Baltimore Sun*.

Illustrated, Cloth, $1.50 at All Booksellers.

LITTLE, BROWN & COMPANY, PUBLISHERS, BOSTON.

could have lost upward of $100,000 if the scammers had not been in too much of a hurry to clean up and clear out.[74]

Pool rooms south and east of Spokane, Washington, and a book operating at the Oakland, California, racetrack, were beaten out of thousands of dollars within the previous few weeks, according to a January, 1907 article. The gang operated by getting the results of the Oakland races to Spokane by direct wire from the coast and them "flashing" them by long-distance telephone to confederates in Butte, Montana, and other cities. By that method, the confederates placed in the inland cities were able to get the results by phone from five to 10 minutes before the results could be relayed by telegraph. Pool rooms at Butte also were hit with large losses. A spokesman for the bookies admitted, "as the messages from Oakland to Butte have to be relayed at Ogden, Utah, and Helena, Mont.," the gang had an extra five to 15 minutes to work with. Spokesman I. L. Hildebrandt added, "The Spokane and Butte men turning the trick are well known to our men. They know we are wise; also that we can't touch them, because they are outside the pale of the law."[75]

Yonkers, New York, police arrested six men on August 30, 1907, all of whom had furnished fake news of the races to bookie joints. Two, Albert Hopper and Frank Gill, were from New York City, while the other four were from Yonkers. Allegedly, the men rented a house that overlooked the racetrack. From there the news was sent to Alberson's ice house, in a hollow below the track, from which it was sent to some other station of wireless telegraphy. Police raided the house at 3:00 p.m. on August 30 as news of the second race was being sent out, and arrested all six men. They were held for violating Section 639 of the New York State Penal Code as they were alleged to have tapped Alberson's telephone wire. Apparently these men used a different sort of system that involved physically observing the race and then sending the results on through a faster wire system that reached its objective much more quickly than did the regular wire message from the track to its pool room customers. However, the system was not explained in any coherent fashion. "Wireless telegraphy" was just starting to come into its own at this time. While the start of the trial got a brief mention several weeks after the arrests, no more reports on the outcome were published.[76]

The days of the wiretappers who scammed the bookies were coming to an end. It was getting harder and harder to operate a successful fraud. More and more precautions were being taken by telegraph companies and, more importantly, telecommunications were changing. The days of sending out race results by Western Union wire, and taking up to seven minutes to transmit such a message (coast to coast) were in the process of being usurped by such devices as the long-distance telephone and wireless telegraphy—that is, radio. Conventional wiretapping of telegraph wires for racing scams, which had been going on for some 25 years, had only about three more years or so of life left.

Out in Santa Rosa, California, in April 1908, one of the two local pool rooms was scammed out of $400 by Walter Rea, a "well known young man about town," observed a journalist. He bet $5 on an 80-to-one shot. When word was received his horse had won he cashed his ticket and left town. It was thought that a confederate tapped the wire and sent in the wrong horse as the winner. When official results were received at the pool room the fraud was discovered. Rea was arrested on a charge of grand larceny. He was released on $1,000 bail. No more reports were published on Rea.[77]

T. C. Diffenbacher was confined in the City Prison in San Francisco in April 1908, pending investigation. There was no specific charge against him. Diffenbacher admitted he was a wiretapper and had beaten pool rooms out of large sums of money but he claimed that his business in San Francisco was a legitimate one. He said, "I came to California for my health but frankly I would have done a job here if the opportunity had presented itself." The police were reported to be in a quandary about what to do with the man as he had committed no offense against the laws of California since his arrival in the state. Two days after he was taken into custody, Diffenbacher was released, as the police could not find anything to charge him with. However, noted a news report, "His outfit of wire tapping instruments is being retained by the police."[78]

Two private detectives employed by the Pinkerton Agency were said to have nipped in the bud a gigantic wiretapping scheme by arresting two men on April 22, 1908, near the outskirts of Vallejo, California. The men had a complete wiretapping apparatus in their possession and were in the process of installing it. Arrested were T. B. Hammond and W. G. Morton. Reportedly, the Pinkerton men followed the alleged wiretappers from Portland, Oregon, when they left that city two days earlier. The quartet arrived in San Francisco on the morning of April 22 and immediately took the boat to Vallejo. After spending a few hours around town, the men rented a buggy and drove to a secluded spot on a country road. The detectives watched for a time and then swooped down on them when the men had almost completed the work of tapping the wire. Telegraphic instruments were found in their buggy along with other items and tools used by wiretappers. Nothing more was reported on this case.[79]

Wiretappers operating in Tacoma, Washington, in May 1908 managed to make off with over $2,500 from the Warwick Turf Exchange before their fraud operation was discovered. Three men were suspected but by the time the discovery of the scam was made they had already left town. No report of the fraud was made to the police and the owners of the bookie joint were reported to be trying "to suppress all details of the affair." They had perpetrated the usual type of fraud by holding back the race results from the telegraph system and placing their winning bets in the 15 or so minutes of delay time created by holding back the results. By occasionally losing small amounts on horses that they knew had already lost, and by scattering their bets, the gang succeeded in keeping their operation hidden for a week or more. It was reported that private detectives had been engaged by the Warwick Turf Exchange owners to watch for the crooks, should they happen to return.[80]

A June 1908 article observed that in New York City the Jockey Club and the racetracks had resumed their war against the pool rooms in earnest. A censor was placed in charge of the Western Union telegraph wires inside of the Sheepshead Bay track and nothing but newspaper reports, delayed, were sent out during the afternoon. (Telegraph companies had always sent race results to the newspapers.) It had been decided not to completely remove the wires from the tracks as had been originally proposed, but the censorship at the track was to be very strict so that, it was said, even private messages would be held up. By that arrangement, it was impossible to send any information whatsoever to the New York City pool rooms, which had been getting "almost perfect service" since the new betting law went into effect. The censorship move was undertaken by the track owners and the Pinkerton detective agency.[81]

Officials at the Pacific State Telephone and Telegraph Company declared on January 22, 1909, that they had broken up a gang of wiretappers. Reportedly it was headed by Charles H. Adams, alias J. H. Schneider, whose reputation as a wiretapper, said a journalist, "is known all over the country." Adams graduated from Harvard University and subsequently was appointed a lecturer at Cornell University. Other members of the gang were Charles Beander and Robert Moore. After serving a term in jail in the East for forgery, Adams, 43, came to the West Coast and had commenced wiretapping in Vallejo about eight months earlier but no direct evidence was obtained against him. He was charged with a misdemeanor and allowed to go on a promise to leave the state. However, he went to San Francisco in the autumn of 1908, at which time the telegraph company became aware that wiretappers were at work on one of their racetrack lines. When this work was detected Adams fled the area. He next started tapping at Visitacion Valley in San Mateo County. On October 6, 1908, an official with the telegraph company found the three men at work on his firm's lines. They were arrested. Adams was tried, convicted of wiretapping on December 15, 1908, and sentenced by Judge Buck on January 5, 1909, to serve a short term in San Quentin. At that point Beander had been tried and convicted but not yet sentenced. Moore had not yet been tried. No more reports on the outcomes were published.[82]

Joseph H. Robinson was tried in Oakland on June 3, 1909. The judge in the case ruled that unless it was shown that someone had been defrauded, tapping a telephone or telegraph wire was not a crime. In the case at hand Robinson had never actually used the information on the horse races to bet in the pool rooms, and the charge against him was dropped.[83]

Reportedly, between $50,000 and $70,000 was taken in Denver and Salt Lake City in November 1909 as the result of a wiretapping of the Latonia racetrack results. It was said that nearly every bookmaker in and around Denver was hit and that the operation extended as far as Chicago. Those who controlled the wires declared they would launch an investigation but no mention was made of the police being called in.[84]

During the early part of April 1901, it was reported that a huge wiretapping scheme had been pulled off in the pool rooms of Philadelphia. It was an operation that extended to New York City, Chicago, and many other large cities. A racetrack in Oakland, California, was the one tapped. Reportedly, $100,000 had been taken in Philadelphia but, it was said, up to $1,000,000 had been fraudulently obtained nationwide. Confederates in pool rooms around the nation had bet on earlier races throughout that day in order to avert suspicion. A later report declared the amount taken by the scammers might have been "exaggerated." Once again, a thorough examination was said to be in process by officials of the telegraph company and, again, no mention was made of police involvement.[85]

A gang of five men from Chicago claimed they had scammed a total of about $85,000 over a period of about three weeks, in February 1911, from a string of New York City pool rooms by tapping the Jacksonville, Florida, race wire just across the Hudson River. Two of the five men stayed at one hotel while the other three used a different one. By playing the horses at various bookie joints they gradually made friends and met people. At one place they chanced upon a telegraph operator from whom they learned what they needed to know. The same man never worked the same pool room twice. There was no indication the police were called in once the fraud was discovered.[86]

5

Wiretapping the Markets: Stocks, Commodities and Bucket Shops

A newspaper account from May 2, 1878, gave brief and sketchy details about some of the wiretapping that took place in that decade with respect to stocks. In the days of the lawsuit between the Chollar-Potosi and Burning Moscow mines (around 1873–1878), a telegraph operator was sent from San Francisco by some of the mining magnates to tap the wires at Sportsman Hall, a locality about 12 miles away from Placerville, California. In those days the only telegraphic communication with Virginia City was by this route. That operator was detected, arrested and finally sent to prison for a term of two years. In 1878 Nevada Senator William Sharon (1875–1881) complained that telegraphic messages sent to him in code had been deciphered and the contents sent to others who speculated in stocks. A few years earlier a millionaire had made arrangements with an operator in Virginia City to decipher the dispatches to Senator Sharon, especially those relating to the Chollar mine. Both times, Sharon was reportedly surprised and indignant. In that earlier instance the operator was detected and, when confronted with the evidence, admitted he was the one who had deciphered the dispatches for Sharon and forwarded them to a broker located in San Francisco. As well, there were said to be authenticated stories of how a man named Hayward's dispatches from Crown Point were deciphered; several well-known brokers were connected with that tampering.[1]

Warrants were issued at the request of the Western Union Telegraph Company in Chicago on October 17, 1883, for the arrest of C. F. Van Winkle and William Alkorn on a charge of conspiracy. The story began on May of that year when Western Union refused to furnish market quotations to the bucket shops and removed the telegraph tickers that had been installed therein. As a consequence, the bucket shops had been working all sorts of schemes to secure stock reports and quotations, with varying degrees of success. For a time it was supposed that they managed to receive quotations by means of signs and signals from persons in the Chicago Board of Trade (established in 1848 and the world's oldest futures and options exchange). Then it became known that they were receiving quotations just as they were sent over the Western Union wires. Upon investigation it was found, allegedly, that Van Winkle and Alkorn had tapped the telegraph wire and had run a branch wire into a small office in Chicago where the quotations were taken off and sent to all of the different bucket shops in the area. Van Winkle was then arrested and held in $1,000 bail while Alkorn had not been located. Bucket shops flourished in the United States from roughly the 1870s to the 1920s. They were effectively eliminated in the United States around 1920 by laws passed against them everywhere, as they were largely viewed as scams. A bucket shop establishment nominally existed for the transaction of stock market business or business of a similar nature, but really existed for the registration of bets or wagers, usually for small amounts on the rise or fall of the price of stocks, grain, oil, or other commodities. There was almost never a physical transfer or delivery of the stock in the form of a certificate or any of the commodities nominally dealt in. A bucket shop was a place for side bets—like a bookie joint—but not a boiler room, which tried to sell low-value and essentially worthless stocks to unwitting people. A legitimate stock broker was duly registered in this time period and had the ability to buy and sell stocks, receive the certificates and transfer them to his clients, and so forth. A bucket shop was on the outside and not registered and could not buy stock on the exchange and take delivery of the physical certificate. On rare occasions a bucket shop did engage in such trades in the hopes that it could show such transactions to the authorities as "proof" it was not some sort of scam. When it did so it had to buy through a registered broker and since those brokers did not deal with bucket shops, it had to do the deal clandestinely.[2]

Sid Phelen was a resident of St. Louis and had cotton exchanges in Atlanta and the Alabama cities of Birmingham and Montgomery. He returned to his home town in March 1887 after having trouble in his Birmingham exchange with wiretappers, an experience that was said to have cost him about $10,000. Market quotations were sent from Atlanta via Montgomery to Birmingham with each exchange receiving the news at the same time. If the wire over which the news to an exchange traveled was tapped, the quotations could be manipulated and trades could be made in the exchange to suit the tappers. Being in possession of the telegraph line meant the tappers had the power to send the market either up or down. About two months earlier, said Phelen, the wire leading to the Birmingham exchange was successfully tapped somewhere inside the city limits and the manipulators soon struck him for $18,000, though he had to pay only $8,000 of that amount. Phelen remarked, with respect to the tappers, "Why, they had a dead sure thing, as they were fixing their buying and selling price, and we were dancing to their music." When the manager of the Birmingham exchange had his suspicions aroused he asked for price verification. That request had to go to New York and the quotation had to come back signed by the manager of the Gold and Stock Telegraph Company. When that request was sent in a reply was received in about 20 minutes. However, it came from the tapper, who called up the exchange and told them the quotation was okay, signing himself with the appropriate name. Those manipulators had such complete possession of the quotations that Phelen's Birmingham office could not get the correct figures from his other exchanges, except by mail. As soon as the quotations from Atlanta were compared with the quotations in Birmingham the fraud was discovered and in settling up Phelen refused to pay $10,000 of the fraudulently obtained gains. The wires were tapped for a second time at the Birmingham exchange on March 1, 1887, with the fraud being done in a different fashion. This time the quotations were not manipulated but were taken off the wire and held back until trades could be made on the basis of quotations known only to the tappers. On the morning of March 1 Birmingham asked the Montgomery exchange for the time of day. That was intercepted by the tappers who sent them a time making their clock 12 minutes slow. As the quotations came in the tappers took them off and held them back for 12 minutes. Thus, they knew the real prices 12 minutes ahead of the Birmingham exchange and could trade accordingly, either buying or selling. The loss to Phelen was about $4,000. However, when it came time to settle up, Phelen sent the matter to arbitration and the order was made that Phelen should pay only $1,600. That same method of arbitration was used in the first case of tapping, resulting in Phelen's payout being reduced to $10,000 from $18,000. He had to pay some of the loss because there was no direct evidence against the perpetrators and so they got away with it, at least to some extent. Phelen told a reporter he would guard against wiretappers in the future by putting in a set of quadruple telegraph instruments, which would supply too much electricity for the wire tappers to be able to manipulate—or so he thought.[3]

A July 1889 news article briefly noted that some telegraphers had made money by "scalping" the oil, grain, and stock markets. When trading was lively the telegraphers employed at the exchanges frequently got quotations from Oil City and Chicago two and three minutes in advance of the official quotations. "At such times a tip to a broker is worth more to the operator than his month's salary."[4]

A May 10, 1893, report from Michigan noted that the public produce and stock exchange of Chicago, with a special wire and branch offices in Dowagiac, Kalamazoo, Grand Rapids, and Muskegon, lost between $1,500 and $2,000 on May 9 through the tapping of its wires. Apparently the wire was tapped near Niles and confederates in the four branch offices made deals in stock that netted them $900 in Muskegon, $400 in Kalamazoo, $120 in Grand Rapids, and an unstated amount in Dowagiac. The fraud was discovered after business hours when the transactions of the day were being checked.[5]

Thomas J. Dunn and Walter Elliott were arrested in the office of H. M. Quinn & Co., brokers, at 50 Broadway in New York City on March 22, 1898, on a warrant issued by Magistrate Cornell charging them with tapping wires in violation of section 639 of the New York State Penal Code. The complaint against them came from Walter Content of the firm H. Content & Co., brokers at the same address. Police said Dunn was arrested in a bucket shop raid several months earlier and at that time gave the name of H. M. Quinn. They declared that Dunn was the head of the Quinn firm. Both men denied any guilt. Content's lawyer told Magistrate Cornell that Content first noticed something on March 17 when he thought something was wrong with his private telephone wires that reached from his office to the New York Stock Exchange. He chanced to come to his office earlier than usual on the following day and found a man working among the wires within the building. That fact, plus the knowledge that some private information sent over the phone the day before had become whispered about on the street, convinced him his wire had been tampered with. He notified the police and the Metropolitan Telephone Company and an investigation was started.

Telephone employee Thomas Carney discovered that phone line had been tapped. By following that wire Carney saw it led into the office of Quinn & Co. He got into that empty office with the help of the cleaning woman. With Carney at the Quinn phone and Content at his own (in his own office) the two men could talk to each other. Three policemen made the same test with the same result. Carney also discovered that whoever tapped the wire had known that it was a private wire and that any difference in the resistance, more or less, would be easily noticed, for extra batteries had been fixed onto the wire so that the tapping would not be betrayed by a decreased volume of sound when messages were being received. With that evidence the police got arrest warrants and picked up Dunn and Elliott. Lawyer Myron Oppenheim told a reporter that a large sum of money had been made by the conspirators at that point. He said a number of times it had been noticed by Content that various people at the Exchange always bought or sold stocks just at the right time to accord with the private information Content & Co. had but a few moments earlier received over their private line. Cornell held the two men on $1,500 bail. Dunn furnished the bond and was released but Elliott was locked up in the Tombs prison.[6]

Panic reportedly occurred on September 29, 1899, on the floor of the New Orleans Cotton Exchange shortly after business opened for the day, and caused the complete suspension of futures business pending, said an article, "the investigation of what at the moment was assumed to be a gigantic conspiracy to swindle the Cotton Exchange of the county." That panic was due to an apparent large jump in the price of cotton, based on alleged Liverpool, England, information; it was estimated that roughly $170,000 had been lost on local cotton transactions as a result. Added the reporter; "Cotton men say wires were tapped as a part of a deep-laid plot to ruin the bucket shops of the country and this opinion seems to be well founded." On that particular day the New York market was closed for a holiday; thus, the Liverpool advices came direct to the New Orleans Exchange instead of by way of New York as was the custom. Supposedly that gave the wiretappers their opportunity. At Liverpool the market opened down in price as compared to the day before and continued that way without any change for some time. Then the wires heated up with tales of rapidly advancing prices. In the meantime New Orleans had opened under the influence of the Liverpool information much higher, compared with the previous day's close. It quickly climbed some 30 more points. With a price rise of 54 points facing them operators began to receive cablegrams from Liverpool asking the reason for the heavy gains in prices in the New Orleans market while stating that prices in the English market remained at their opening figures. In the meantime some 40,000 to 50,000 bales of cotton had been sold in New Orleans at the much higher prices. Finally the New Orleans exchange, by then in a state of panic, suspended business for the rest of the day.[7]

President Parker of the New Orleans Exchange had his counsel Mr. Saunders declare, "Having been informed that all contracts made this day were based on false reports from Liverpool as to price I advise you that the consent essential to a valid contract was wanting, and the contracts are therefore void." All of the Liverpool dispatches came through the regular channel, the Commercial News Bureau of the Western Union Telegraph Company, which had been the news conductor for the various Southern exchanges for more than 25 years.[8]

On the following day it was reported that it was speculated that Cincinnati men were involved in the cotton false cables. That story surfaced when local Cincinnati brokers compared notes. Liverpool quotations were sent in as being up 77 points while cotton was really off 2.5 points on the Liverpool market. Based on those false wires cotton jumped in price by $2 a bale in New Orleans. It was said that a half a dozen men were around at different brokers' offices in Cincinnati early in the week talking about cotton and trying to buy the May option. Those men were described as strangers who "looked like Southerners."[9]

Also one day after the event offices of Western Union had remarked silent about the swindle. An employee of the company did speak anonymously to a reporter and said that Western Union felt it was almost impossible for someone to tap the wires: "I have been told that attached to the switchboards are incandescent lights which indicate the power of the current passing through the wires, and also whether the current is affected at any point." Should a wire be tapped, the current would be broken and the break would cause the current between New York and the former point to rise. That rise would be immediately indicated on the switchboard by the flaring up of the electric lights. That, of course, left Western Union to contemplate the only remaining option, a crooked employee. Whenever Western Union was faced with a situation like this, with fake messages going over the wires, it almost invariably chose to put the blame on one of its employees, even when no evidence existed. Employees were often dismissed and/or suspended from work even on the basis of no evidence, if they seemed to be likely suspects. This tactic was, of course, in the interests of Western Union. An admission that wiretapping had been successfully carried out tended to damn the entire Western Union system, while a rogue employee was just that, a single bad apple and no reflection on the system and its methods of operation.[10]

A prominent member of the Minneapolis Chamber of Commerce received a letter on June 21, 1902, from Kansas City, Missouri, giving the details of the recent exposé of wiretapping at the Kansas City Board of Trade (a commodities futures and options exchange like the Chicago Board of Trade). For a period of some months the Kansas City Board had been puzzled as to how the Christie Grain Company and the John J. McPherson Company got their quotations. Neither firm had any standing on the Kansas City exchange, or on any other exchange. (They were bucket shops.) Every possible effort had been made to discover the leak but no success had been achieved. One day John T. Snodgrass of the quotation committee received an anonymous letter suggesting he examine the duplex wires running from the tower of the exchange building to the Gibraltar Building. Snodgrass employed electricians but they found nothing wrong. He continued to investigate and finally became convinced that the bucket shops were getting their quotations from the F. P. Smith Company on the fourth floor of the Exchange building. As the Smith people held "first-class" standing, Snodgrass hesitated about bringing so serious a charge against them. Finally he reported the matter to the Chicago Board of Trade for further investigation. It happened that the Smith people had the private wire of Harris, Gates & Company of Chicago, of which firm John W. Gates was a member, so the implication of connivance at bucket shop operations rested alike on both and the honor of the Harris, Gates firm was at stake. Determined to clear his firm at any cost, A. H. Farnum of Harris, Gates went to Kansas City. With him were two expert linemen. They spent 10 days tracing the maze of wires in the Exchange building. Finally they landed in the office of F. P. Smith, which firm was entirely unaware of any leak in their office. One of the investigators went to a blackboard and began to chop into a wall with an ax. After a time a niche was exposed wherein was found a miniature telephone transmitter so placed that it could catch the click of the telegraph instrument that was located near the blackboard.[11]

The evidence in the Kansas City case was conclusive: the bucket shops were stealing the quotations of the exchange. Since the courts had decided earlier that such quotations were property it meant the eavesdroppers were stealing the property of the exchange. With that as a basis it should have been a simple matter to punish the perpetrators through the courts; yet a newsman pointed out that technicalities stood in the way. While a suit for wiretapping might have been brought against the perpetrators, "it was the opinion of eminent counsel that while the bucket shops were certainly stealing the quotations it would be doubtful whether the court would construe the placing of telephone transmitters behind a blackboard as wire tapping in the strict sense, hence no action was taken." Farnum looked forward hopefully to the end of bucket shops through the various court rulings around the country that went against them, for reasons other than wiretapping, and predicted their end. A growing hostility toward them was evident, Farnum argued, and everywhere increasing legal difficulties with respect to their operations were put in front of them.[12]

After some investigative work the Minneapolis Chamber of Commerce directors announced in October 1902 that they had discovered the leak by which bucket shops had been securing quotations on the local wheat market. The price clerk was screened and was found to be doing nothing amiss. But it was still true that Chamber quotations were still being received at the bucket shops. That left one possible source to check—the telephone boys. Ranged around the walls of the big exchange room were 22 telephones, maintained by area brokers who used them for sending quotations from the floor back to their offices. Each phone was in the charge of a single boy. After notifying the spotters in the bucket shops of their plans the directors began taking the boys away from their instruments at intervals of 15 minutes. That procedure went on for an hour or two with no results. Suddenly the spotters reported a break in the quotations coming to the bucket shops: nothing was being received. That fixed the responsibility upon a boy employed by a Minneapolis broker to send quotations to a St. Paul correspondent. Both were in good standing with the Chamber and the directors were hesitant to accuse either one or both. Wiretapping was regarded as the logical explanation but since the leak was then stopped, said a reporter, "an embarrassing inquiry may not be pushed further." At the same time the Minneapolis Chamber of Commerce and the Western Union Telegraph Company had agreed to end a fight between them. Western Union had agreed to sign the contract demanded by the Chamber to withhold the Minneapolis quotations from the bucket shop provided the Chamber would sign a bond to indemnify Western Union from any damages that might result from such actions—such as a lawsuit from the bucket shops. In this case, as in many others, the telegraph company took the position that it was a common carrier (as it had been so cited in numerous court decisions) and as such could not discriminate as to what firm or firms it would supply with continuous quotations any more than a railroad could discriminate against any passenger or refuse any freight, also under the common carrier doctrine. The Chamber argued it had moral right on its side as Western Union was taking the quotations from the Minneapolis Chamber of Commerce (its best customer in the Northwest) and selling them to bucket shops just to make money.[13]

A Postal Telegraph Company operator was arrested in Memphis, Tennessee, on January 3, 1906, on a warrant sworn out by Mr. Bryan, manager of the Postal firm, charging him with wiretapping. He cut in on the main Southern wire of Logan, Bryan & Company of New York and Chicago, using a vacant room in the Postal building. After obtaining the information he was looking for he phoned those quotations to a Memphis brokerage firm. The crime of wiretapping was only a misdemeanor under Tennessee law but the firm that received the stolen quotations "will probably be severely dealt with," said an account. "That they are implicated is shown by the fact that the two members of the firm signed the bond of the prisoner."[14]

On July 19, 1906, in St. Louis, Henry Stanley, a telegraph lineman, was arrested on the charge of being implicated in a wiretapping scheme. The police believed they had brought to light an extensive scheme for the stealing of quotations from the merchants' exchange in St. Louis and that wiretappers had been operating in St. Louis for about three months. However, Stanley was the only one arrested to that point.[15]

A long article published on June 3, 1908, discussed at length the continuing demise and near death of bucket shops. It argued that the obliteration of such places would lead to the people of the United States saving an estimated $200,000,000 annually. For years the bucket shops had shrugged off the war made upon them by the legitimate stock and commodity exchanges that they discredited by their very existence. But at that point, said the piece, there were strong and successful attacks being waged against them all over the country. In May 34 indictments were handed down against them in Cincinnati while on May 19 in Los Angeles every bucket shop in the city was raided and proprietors and customers were put in jail. Oklahoma had just signed into law an anti-bucket shop bill that went into immediate effect. By that very same night all of them in the state were driven out of business. Prosecutions were then under way in St. Paul, Minnesota against the Superior Grain Exchange and the Wisconsin Grain and Stock Company. Missouri followed and put almost all of St. Louis's bucket ships out of business after a law was enacted against them. The largest bucket shop operation in America was said to be Morehead & Company, which had 300 offices scattered around the nation and linked together by private telegraph wire. The next two largest were the George H. Stapley Company and Brown & Company (200 offices in total for the pair). Stapley had just gone into receivership. Those three firms, declared the reporter, were now almost completely removed from the landscape. "A bucketshop is a gambling device with all the paraphernalia of a brokerage house. The transactions are wagers, mere gambling under the guise of commercial transactions," he argued. The bucket shops made their profits from the losses of their customers, unlike the legitimate brokerage houses that earned their profits from commissions. It was a common practice for the big bucket shop concerns, he added, to maintain exchanges where employees pretended to be brokers "buying and selling" with an exact imitation of every small detail of trading on the legitimate exchanges. For example, he continued, with respect to the Superior Grain Exchange, "all the quotations were stolen from the Chicago Board of Trade by means of wire tapping, three-eighths or one-fourth of a cent added to them and sent out as quotations of the Superior Grain Exchange." In that manner firms against which the Chicago Board of Trade had secured injunctions restraining them from using their quotations were able to evade the law and secure the necessary figures. The instant the bucket shops had no quotations to write on the blackboards, the betting in such places ceased. The United States Supreme Court had held, concluded the newsman, that the Chicago Board of Trade was not, as bucket shops alleged, a gambling institution and that it had a property right to its quotations.[16]

The above article prompted an editorial on the subject the next day from an Arizona newspaper. That state had recently signed and put into effect a law against bucket shops. As well, the U.S. Post Office was depriving them of the use of the mails, as fraudsters. "The legitimate exchanges have an added reason for fighting the bucketshops for the latter have made it a habit to steal the quotations of the former by wire tapping," declared the editor. He believed the public was often wary of the stock exchanges in general by confusing the operations of legitimate brokers and those of bucket shops. "The method of the latter is simplicity itself. It charges a commission, usually double that of the legitimate broker, and except in rare cases executes none of the customers' orders. It is a gambling enterprise out and out with the heavy commission always working for the bucketshop," concluded the editor.[17]

E. B. Saylor, former superintendent of the Pittsburgh district for the Western Union Telegraph Company, was arrested on May 15, 1909, on a charge of misdemeanor in connection with an alleged wiretapping scandal. He furnished the required $1,500 bail. Three other men were also arrested: brokers William L. Thompson and William H. Smith and Isaac N. Barto (a Western Union employee), all charged with misdemeanors. H. A. Foss, representing the Chicago Board of Trade, alleged that on April 26, 27, 28, 29, and other days prior to that, the defendants conspired to cheat or defraud the Chicago Broad of Trade of certain quotations on grain by tapping wires carrying the quotations and furnishing the same to various persons.[18]

At a hearing in Pittsburgh on May 18 it became apparent that the operation had been going on for about five years. Barto turned state's evidence and cooperated with authorities. He told how he and other employees of Western Union had received money from local bucket shop operations to obtain quotations. Several detectives from New York had been employed earlier to investigate and followed Saylor and Barto to the hotel where they met Smith and Thompson, with the former receiving an envelope from the latter. After that Saylor reportedly gave Barto $75, which he split with a couple of other Western Union employees.[19]

During the trial of Saylor and Smith, Barto testified that at one point

Saylor had advised him to flee the area and to go to Canada. H. A. Foss told the court that as early as February 15 the Board had found its quotations posted on the boards of Pittsburgh bucket shops before their receipt by lawful customers in Pittsburgh. On September 13, 1909, Saylor and Smith were found not guilty by a jury on a charge of conspiracy in connection with the Chicago Board of Trade wiretapping case. Thompson, also indicted but not on trial, was ordered acquitted by the court.[20]

Charges that he had conspired to furnish stock quotations sent from Chicago and New York City over private wires of the Western Union Telegraph Company to persons not entitled to receive them were made against John P. Altberger, superintendent of the Philadelphia Western Union Telegraph Company on September 7, 1909, when he returned to Philadelphia on that day after a month's holiday in Europe. He was placed under $5,000 bail. It was reported that "the prosecution is backed by the Western Union company." Altberger denied all the charges. It was understood his arrest followed an investigation made a few weeks earlier while Altberger was abroad with his family. For some time, Western Union officials said, wires over which cotton quotations and quotations from the Chicago Board of Trade were sent had been tapped by bucket shops and brokers not connected with those commercial bodies. Altberger was eventually indicted and set for a trial but then he got caught up in something bigger when his name was one of some 40-odd cited when Western Union itself was indicted for running a bucket shop in Washington, D.C. And then it all just faded away and disappeared from sight.[21]

An editorial appeared in a newspaper in Washington, D.C., on April 3, 1910 in praise of the U.S. government's move the day before to launch a crusade against bucket shops all over America. Raids were made of such establishments in every large city in the country, after indictments had been secretly returned by the Federal grand jury in Washington. Reputable stockbrokers had, reportedly, endorsed the move. Evidence against the bucket shops had been gathered from many sources including the U.S. Post Office and the U.S. Department of Justice. "In some cases wiretapping was resorted to. No matter how gained, the evidence is now in the possession of the government, and it will be used to the best advantage," noted the editor. Arrests had been made in many of those cities as a result of the raids and the editor thought the blow to the bucket shops from this latest crusade would be a fatal one.[22]

Special agents of the Department of Justice had been carrying on the nationwide campaign against bucket shops quietly for about two months. For an even longer period they had received numerous complaints from people who had been swindled by bucket shops. In the eastern part of the country three concerns were said to have controlled most of the business: E. S. Boggs & Company of New York and Philadelphia, Price & Company of Baltimore, and the Standard Stock and Grain Company of Jersey City. Boggs had 101 branch offices, Price 81, and Standard 76. Each of the three firms paid the Western Union Telegraph Company about $100,000 in tolls annually for their special wires. Price & Company had 11,000 miles of leased Western Union telegraph wires. It was reported that evidence was hard to get until finally the special agents conceived, noted a newsman, "the idea of beating the devil at his own game. They employed a number of expert wire mechanics to tap the wires of the companies for them, and for days they made verbatim accounts of everything that passed over them." He added, "This was the move that the agents believe won the game for them and which resulted in giving them evidence upon which the federal grand jury in Washington handed down twenty-nine indictments, with more to come."[23]

The bucket shops had two wires on the go. One was a regular stock ticker that printed the New York Stock Exchange quotations as they occurred. The other was a special wire from New York into their offices that could beat the ticker service by 15 minutes. If a customer entered a bucket shop and placed an order for 10 shares of Southern Pacific at 120 and they discovered it had gone down they bought it. If it had gone up they pretended to take the order but later told the customer they could not execute the order. Those bucket shops also bought and sold for a customer on two percent margin, something no reputable brokerage firm would do. Orders were usually never physically executed, unless the customer insisted on the stock certificate being delivered to him. Under that circumstance the bucket shop got a regular stock exchange firm to execute the order. At the same time they charged customers interest on the money they were supposedly borrowing to execute margin orders. By tapping the wires the federal agents succeeded in getting absolute records, for every purchase had to be wired in to the main offices from the branches. In the case of the Boggs firm, a daily business of from 20,000 to 50,000 shares was transacted, with 99 percent of those trades being in ten-share lots. That meant 2,000 to 5,000 transactions were handled daily and yet the records showed actual transactions of fewer than 100 daily, clearly showing the others were "bucketed." That is, it was only in those fewer than 100 daily transactions that a physical stock certificate was involved and changed hands.[24]

A day later it was reported that the Justice Department campaign against the bucket shops was not over, but had just begun. A total of 16 men had been arrested in those raids in New York City, Philadelphia, Baltimore and Washington, and more arrests were expected. As well, it was reported that the Department of Justice had gone to the extent of contemplating action against the Western Union Telegraph Company. The government was said to believe Western Union knowingly furnished the offending bucket shops with tickers and other wire services. That, in the mind of some government officials, made Western Union as much of a conspirator as any of the men arrested. On the other hand, speculated a reporter the department might allow Western Union "to go free with a severe reprimand." It was also noted that reputable brokerage firms charged customers one-eighth of a percent commission while the bucket shops took 1.75 percent. While federal agents had the wires tapped other federal agents went into various bucket shops to buy and sell, and receiving receipts for such transactions. However, the agents doing the tapping reported those transactions did not pass through the wire to Philadelphia to the exchange involved. That is, no trade had really been executed.[25]

A federal inquiry before the grand jury in Washington at the beginning of May 1910 looked into the establishment of a direct connection with the Marrin "fast wire service" from one of the leading wire concerns of the country. That inquiry was looking into the methods of obtaining stock quotations used by bucket shops. Five officials of Western Union (one the previously mentioned Saylor) gave evidence, as did Frank Maier and Thomas Marrin, owners of the service. Testimony was given in secret. Prosecutors hoped to show that the fast wire service was furnished to a number of bucket shops ahead of the regular ticker service. Through the operators the government hoped to satisfy the grand jury that the information came from a specific source and was not the result of wiretapping. The reported discovery of a leak of New York Stock Exchange quotations to bucket shops through a member of the Exchange was denied by the Exchange.[26]

On July 7, 1910, the Western Union Telegraph Company, without any prior notice, cut off the wire service they provided to "unconnected" brokers. That action was taken in Pennsylvania, New York State, and several other states east of the Mississippi River. Few of those places that lost service were openly run as bucket shops, but sailed along under the guise of brokerage houses and maintained tickers and the usual paraphernalia of the legitimate brokerage houses. That sudden stop in service was said to have been the outgrowth of a conference held recently in Washington, D.C., between government officials and Western Union managers and directors. That company was indicted on June 10 for alleged violations of the laws in the District of Columbia against bucket shops. In the indictments were mentioned 42 overt acts. And that led to the conference. A finding of the federal grand jury that had probed the situation was that

Western Union had been connected with the operations of the men charged with keeping bucket shops in violation of the law. The first definite step against the company was taken by the government on April 30 when it put the Marrin Service, which had been the bucket shop feeder, out of business. It was found that the Marrin Service was obtained by Marrin from Western Union. Then there was the major raid launched by federal agents that put many of the bucket shops out of business, at least for the time being. "It was said that recently in some mysterious way the service was continued, and that there was a supposition that there was wiretapping somewhere," observed a journalist. Examples of wire service cut-offs could be seen in Utica, New York, where the Western Union telegraph manager was directed by the company to discontinue wire service to two stock brokerage offices in that city; the wires were then cut. Both of the brokers were agents for a Baltimore company and were doing a large business. Several other brokers were treated the same way in other New York State cities, including Auburn, Gloversville, Rome, Watertown, Ithaca, and Syracuse. However, in Baltimore that day, July 7, Western Union was enjoined by Judge Stockbridge from cutting off the wire in the offices of E. Herford & Company, stock brokers. It was an issue said to be shrouded in mystery as neither side would offer an explanation. The suit alleged that the brokerage firm paid Western Union $7,526 in advance for private wire and ticker service for the month of July with the service being supplied through to the 7th. Then an attempt was made by Western Union to cut off the service.[27]

DeWitt B. Lowe, a well-known Salt Lake City commission broker, and four other men were arrested at 11:30 a.m. on September 1, 1911, on charges of stealing stock quotations and stock information coming from New York over the private wire of Badger Brothers, stock brokers of the same city. They were alleged to have learned the contents of messages going over the Badger wires and to have sold the information so secured to San Francisco brokers. Arrested were Lowe, of the Lowe Brokerage Company; E. F. May, until recently an employee of the Bell Telephone Company; Ray M. Perkins, chief installer for the Bell Telephone Company; R. L. Scott; and J. F. McAllister. In this case no actual wiretapping was alleged. Rather, the work was done by telephone and telegraph contrivances. The record made by the telegraph sounder in the Badger offices was said to have been transmitted to an office in the Walker Building and from there relayed to the office of the Lowe firm, and then in turn to the office of Moss & Company in San Francisco. Those arrests came after a month-long investigation carried out by the Pinkerton Detective Agency. Several Pinkerton agents had been involved.[28]

Some five weeks before the arrests, Badger Brothers became convinced that information intended only for their use was being secured in some manner by others. The Badger Brothers were the Salt Lake City representative of E. F. Hutton & Company of New York City (members of the New York Stock Exchange). E. F. Hutton operated "at heavy expense" a leased wire from New York to San Francisco and Los Angeles, via Salt Lake City. Over that telegraph wire Badger received New York and Chicago stock quotations and other market information. Despite the worry by Badger over a suspected leak, their investigations revealed nothing. So the matter was turned over to the Pinkertons. Their investigation showed it was possible to stand at the rear of the Badger office near a light well of the Boyd Park Building and hear the telegraph sounder in the office of the Badger Brothers. Two of the offices on that floor around the light well were rented by E. F. May, who represented that he was the district manager for H. H. Mosher & Company, San Francisco brokers. Pinkerton agents observed the occupants. According to them, J. F. McAllister, a telegraph operator, was working with a megaphone attachment to his ear and writing out messages that he intercepted from the Badger wire. May, who was seated near him, was telephoning those messages to R. L. Scott, a telegraph operator who was in an office at the rear of the Lowe brokerage company. Shortly after the Pinkerton men had rented the adjoining office to May, and before the evidence against the accused men was complete, the offices occupied by E. F. May were vacated. May then secured an office in the Walker Building. The Pinkerton investigators did not know how the information was still getting through to that new location, but they were convinced it was getting through. Then they discovered that a private telephone, by a circuitous route, connected the room in the Walker Building to a room at the rear of the office of the Lowe firm. A man was positioned with a telephone testing set. Then Pinkerton men heard May repeat over the phone to R. L. Scott, in the rear of the Lowe building, messages which came originally over the Badger wire. Investigators in the Lowe office were said to have heard Scott repeat these messages on a wire connected with a sounder at the blackboard in the Lowe office. From that sounder the man who marked the board secured the quotations. At the same time Scott was alleged to have been sending this information on to San Francisco. After much more work they traced the wire connection from the Badger Brothers to Lowe. A vibrator had been attached to the sounder to greatly intensify the sound.[29]

A couple of days later T. S. Minot, described as a prominent San Francisco attorney, arrived in Salt Lake City to represent the interests of Moss & Company of San Francisco. He declared he would go to court as soon as possible to argue a conspiracy in restraint of trade was in effect and to compel by injunction the Western Union Telegraph Company to furnish quotations as a common carrier to all parties alike. He also said that Western Union, E. F. Hutton, the New York Stock Exchange, and the Chicago Board of Trade would be investigated concerning their alleged violation of the Sherman anti-trust law on the allegations of unjust discrimination in the delivery of stock quotations. Minot added that for many years there had been strife among the New York Stock Exchange, the Chicago Board of Trade and all brokers who were not members of those two organizations. He added that the two organizations had endeavored through the instrumentality of Western Union "to ostracize every competitor, ostensibly claiming that it was in the interest of legitimate monopoly in stock brokerage transactions, but in reality to maintain and control the current prices of stocks in the market to the detriment of the producer as well as the purchaser." Minot continued by declaring, "The Supreme Court as well as various United States circuit courts have held that 95 percent of all transactions on these two stock exchanges are nothing more or less than bucketshop transactions, with no actual delivery contemplated." He saw the dispute as being between two firms, Moss & Company and E. F. Hutton, with the former being on the outside while the latter was a member of those exchanges and a participant in the monopoly to control quotations. As far as Minot was concerned, since no wires had been technically tapped, once the stock information was ticked off of a telegraph instrument in the office of the Badger Brothers "it becomes public property. It floats out of the window and is taken up in the air. Anybody can listen to it just as anybody can listen to the booming of a cannon."[30]

At the end of September 1911 the county attorney's office in the Lowe case issued a new complaint against the five accused, that of fraudulently reading and learning the contents of a telegraphic message being communicated over a Western Union wire from Chicago to San Francisco. In the original indictment, wiretapping had been one of the charges. But a defense motion to dismiss it was put forward on the ground that the law against wiretapping was to protect the public and not to protect private concerns. The accused had placed a telephone transmitter (with vibrator) so arranged against the wall in the Badger office so as to intercept and communicate telegraph messages. It was thought the new wording of the charge might forestall the defendants' motion to dismiss. The judge had not yet ruled on it.[31]

DeWitt Lowe filed suit in district court on February 21, 1913, in Salt Lake City against the Badger brothers (Rodney T., Jesse T., George, and Ralph) for $50,000 damages. The case against Lowe had been dismissed by Judge F. C. Loofbourow several weeks earlier on the motion of the district attorney for lack of evidence. E .F. May, R. M. Perkins, R. S. Scott, and J. F. McAllister were still awaiting trial on the same wiretapping charge. In his suit, Lowe claimed his reputation was ruined, that he was held up to public scorn as a criminal, and so forth.³²

Wiretapping by telephone was the charge in Los Angeles in January 1914 against H. V. Warnock, an alleged bucket shop operator then on trial in Superior Court in that city. He was said to have obtained New York stock quotations sent by private wire from New York to Los Angeles by simply listening on his telephone, the wire to which was carried in the same cable that contained the telegraph wire over which the quotations were transmitted.³³

According to an article in a New York City newspaper on March 16, 1914, the bucket shops had been driven out of Wall Street, but people who wished to speculate rapidly could try the "English system of settlements." Three years earlier, when the so-called fast wires on which the bucket shops depended were cut and numerous arrests were made, the bucket shop business had been declared dead. For a time the operators lay low. But at the time of this article the New York Stock Exchange was reported to be investigating, through William Bishop, a new scheme in Connecticut. Bishop was the head of investigations for the New York Stock Exchange. The fast wire was known to all of Wall Street. All of the quotations of the exchange were gathered by boys on the floor and turned over to a telegraph company. They were then given to the ticker and otherwise distributed. The ticker was reasonably fast, but if the wire carrying the quotations to the ticker was tapped, a clever operator working an ordinary Morse wire could beat it by minutes. With that service, a cunning broker receiving a buy or sell order for stocks could beat his client out of a fraction of a point or more, as he would know in advance what the price would be. While tapped wires had been found and cut many times by New York Stock Exchange people, the interruption to the service had never been long. Western Union had always insisted that it had no knowledge of any leak or where the information was leaking out. The reporter who wrote this piece picked what he described as a likely target to investigate more deeply—the New England Securities Purchase and Sales Company in Bridgeport, Connecticut. The general manager of that firm was Richard E. Preusser. Some years earlier Preusser had killed a gambler named Miles McDonald in Albany, New York. He was acquitted on that charge on the ground that he was insane. By 1907 he was the manager of the Manhattan Stock and Grain Dealers Company in Jersey City. That concern was closed by the police when it was declared to be a bucket shop. Preusser said he was too busy to be interviewed by the journalist. However, the newsman did speak to a Mr. Snyder at the company's office, who explained that the company charged its customers no commissions and asked no interest on margin accounts. How, then, was it able to turn a profit? Snyder mentioned the English system of weekly settlements under which the customer (gambler, really) did not wish to take up his stocks. In that case the company charged him a small amount of interest for carrying his stocks over.[34]

6
Wiretapping Other Businesses

A favorite target for early wiretappers, with respect to other businesses, was news agencies. A very brief report published in October 1883 related that the Associated Press had had a false telegraph dispatch from Bradford, Pennsylvania, in regard to the flowing of an oil well, "imposed" upon the agency. That oil well, it was later learned, had not begun to flow at all. No other details were reported.[1]

Newspaper and telegraphic circles were reported to have been astounded by the exposure made by the *Evening Post* of Chicago on January 14, 1895, of the schemes and crimes of small press associations, through the manager of one of them, to get the news of the Associated Press before its publication. The yearly cost of the Associated Press dispatches was reported to be $1,250,000 and to secure portions of that news some of the smaller organizations had resorted to all sort of devices such as clipping cable dispatches from early editions of newspapers and rewriting them to make them seem to be original, and by tapping the wires leased by the agency for its business. In order to catch those thieves, two of the "best known detectives in the country were employed," said a reporter. It was also necessary to resurrect some old news dispatches printed in Chicago newspapers from eight to 20 years earlier and then place them on the wire that was believed to have been manipulated. That wire was cut outside of Chicago and another one was substituted in its place and used to carry the agency's regular news reports. Reportedly, the trap was well-laid and it resulted in both Pan-American and United Papers getting caught. Within that latter group were the *Detroit Journal* and the *St. Louis Chronicle*. That the planted news did not get further East was due solely to the lateness of the hour when it was sent out on the decoy wire. According to the exposé in the Chicago paper, eight months earlier a prominent railroad man called at the Chicago office of the Associated Press and stated that he had received a request for a press pass from one W. S. Brewer, who signed himself as "General Manager of the Pan American Association." The railroad man said he had never heard of Brewer and wanted some information about him. The letter on which the request for a press pass was typed had the following as its letterhead: "Pan-American News Association, incorporated 1890. General office 162 Washington Street, Chicago; Eastern office, 31 Park Row, New York. Will S. Brewer, president and general manager." The Associated Press people had heard of Brewer. They had known for some time that their news was being stolen from the leased wires. This visit by the railroad official revived the suspicions previously entertained against Brewer and an investigation of his methods of obtaining news was commenced.[2]

After much work a gang of news thieves was uncovered, assisted by a "notorious" wiretapper. Luther Laflin Mills anticipated catching only small fry, but he also caught some of their larger associates. By tapping the wires of the Associated Press, Brewer got his news cheap and therefore could afford to furnish it to others "at low rates." How Brewer got hold of the news was a puzzle until it was learned he had had dealings with the wiretapper. Both he and Brewer were ex–telegraph operators. Mills was entrusted with management of the case while a detective agency was also employed. Meanwhile, the Associated Press did a little detective work on its own. A careful watch was kept on outside newspapers and every stolen dispatch was carefully clipped and pasted to the original copy sent out by the agency. They soon piled up. To gather more evidence, dispatches concerning non-existent events were sent out by Associated Press and promptly appeared in the Pan-American papers. There were also items planted by the agency that had been published back in 1872. At that point there was said to be no longer any doubt as to how Brewer got his news. A detective working on the case made the acquaintance of an employee in Brewer's office and, under the guise of being a Southern fruit land agent, rented desk space at the Pan-American office at 162 Washington Street, Chicago. It took this man some time, but he eventually learned the inside of the scheme; he sent daily reports to Mills. The Pan-American Association, which had since changed its name to the Union Associated Press, and then to the Eastern Associated Press, had its offices in rooms 32, 33, and 34 of the Washington Street building. Room 34 was always kept locked and it was there that the secret work was done. Wiretapping equipment was located therein. Various holes in the wall and desk drawers connected the secret room with the "news" rooms. It was there the stolen messages were taken from compartments and handed to an editor in one of the other rooms to be sent on to the newspapers subscribing to Brewer's agency. Thus many of the regular employees of the agency had no idea what was really going on. Those rooms looked out on the building housing a newspaper that subscribed to Associated Press and it was those wires that were tapped. Some of the employees in Brewer's firm did know the score and would later confess that the news had been stolen.[3]

Then the AP revived an article from the *Daily News* of April 12, 1886, and sent it out. It was an account of an awful cyclone that struck Kansas City that day and demolished an iron bridge, blew many buildings down and killed a number of people. That item was sent out at 12:00 noon over the wire circuit that was being tapped. All the other newspapers (legitimate subscribers) were cut off and did not receive the eight-year-old item as breaking news. Another wire was installed for them. Thus the fake story went to no one but the wiretappers. Following that another fake story went to no one but the tappers. Around 3:00 p.m. the wiretappers were said to have become nervous (for reasons not explained) and sent telegrams all over the county to the papers subscribing to their service warning them not to use the items, that they were untrue. The wiretapper whose work made their news stealing possible was notorious in that line of work. He had been charged with all sorts of offences of that character. "He is the man who, it is said, obtained for the bucket shops the Board of Trade quotations when all telegraphic instruments were excluded from the exchange," wrote a reporter. Brewer had been an employee of the Associated Press many years earlier "but was discharged for irregularities." He induced a wealthy New York man, Alonzo Rothschild, to invest money to become president of his Union Associated Press. Rothschild, it was said, did not know of the scam. Punishment for tapping wires in Illinois was a fine of $300 to $500 or a prison term not to exceed one year. That evening Luther Laflin Mills, a well-known corporation lawyer and orator, declared, "The extent of the evil exposed in the Post newspaper article this evening on wire-tapping cannot be overestimated. It exists throughout the country in an alarming degree, affecting the interests of not only the safety of confidential communications between persons, but on the vast organization for the public dissemination of news." He added, "The situation is a serious one and, while it is prohibited by the local statutes of many of the States, it seems to demand Federal legislation based on the interstate commerce act which shall supplement such statutes, and it is probable Congress will be asked to take action in the premises."[4]

With regard to the above case, an editorial appeared in a Washington, D.C., newspaper. "So long as wire-tapping was an offense committed by one set of rogues to filch from the pockets of others in the same line a share of their immoral gains [meaning bookie joints], the public was not much concerned, but since the misdemeanor has developed into the stealing of news from the wires of a news-gathering and news-distribution association by a competitor in the same business, it has achieved considerable prominence," declared the editor. "It is not strange that such dishonesty should be found among the riff-raff always to be found in some branch or other of the horse-racing and pool-selling business, but it is rather surprising that such conscienceless conduct should develop in connection with the journalistic profession." To deal with such misconduct, continued the editor, Senator William Chandler of New Hampshire had introduced a bill into the U.S. Congress providing that the conviction of any person who should "wrongfully tap or connect a wire with the telegraph or telephone wires of any person, company or association engaged in the transmission of news" would be followed by a punishment of "a fine of not more than $2,000 or imprisonment, not exceeding two years, or by both such fine and imprisonment." As far as this editor was concerned, no one could object to such legislation and he hoped it would "become a law at an early date."[5]

That same newspaper ran a long article on the above case a couple of weeks later. It declared that with the recent introduction of Senator Chandler's bill into the United States Senate to penalize wiretapping, at least as it applied to the theft of news, "the first step was taken to place electric stealing under the ban of the law. At present, and until Senator Chandler's bill becomes a law, the telegraph companies' only recourse is to have the tappers arrested for trespassing." They added, "The penalty under that law is so small that it has little effect upon the evildoers, and as a consequence, attempts to steal electricity are of almost weekly occurrence." Reportedly, the Associated Press revelations had resulted in a movement to prevent the theft of news in the future by wiretapping and the first result of that had been Chandler's bill. It was quite a different matter to steal stock quotations and racing returns from the Western Union Telegraph Company than to arouse the ire of the business news associations, and when the latter was made to feel the full effects of that evil they immediately set to work to stop it, according to the reporter. When the telegraph companies tried to prosecute wiretappers under the trespass laws, those firms generally took refuge behind the idea that the racing news was aiding a gambling scheme. In the hands of a shrewd lawyer their case generally triumphed. The journalist argued that 60 percent of the telegraph operators did not have the knowledge to work such a scheme from start to finish and, if they did attempt it, "make miserable failures." Yet it was from a great part of that class that the idea of wiretapping emanated. They thought they knew enough, but their knowledge of electricity was weak: "This class of telegraphers may have been successful fifteen or twenty years back, but with the improved apparatus they are never successful."[6]

This article then went on to describe the old and the new types of tappers. The old-fashioned, original wiretapper selected as his scene of operation a pole set back as far as possible from the highway or railroad over which the line repairmen passed when they were looking for troubles. If the pole was partially hidden by trees, so much the better because it lessened the chance of being discovered. Waiting until night, the tapper strapped on his climbers, grabbed his tool sack and climbed up. Reaching the cross arm, he produced straps and a vise with serrated teeth, attached to a loop strap eight to ten feet long. Getting hold of the wire he wished to tap, he buckled his strap around the crossarm and pulled in the slack of the wire until he had six to 12 inches to spare. Then with pliers or nippers he cut the wire off short between the vise and the insulator. The wire did not fall because the strap and vise held a firm grip on it. Then the tapper let in a section of non-conductor, such as a piece of gum-covered wire, making the splice outside of the covering so that no connection was made between the line wire and the newly set-in piece. That was the scientific way. If he was in a hurry or short of material he simply took a hitch around the glass insulator of the tie wire. There was, however, no connection between the two ends. The wire was, in telegraphic phraseology, open in either case. Of course, the operator did not allow it to remain open for long. He made what was called a half connection to keep the wire closed while he continued his operations. He had in his tool sack a coil of fine copper wire covered with dark silk so that it was invisible more than a few feet away. One end of that coil he attached to the line wire on one side of the piece of non-conductor he had set in; the other end he connected to the tapped wire at a point beyond the non-conductor. Then he knocked off his half-connection, for the circuit was completed through the coil of copper wire. Dropping the coil of wire to the ground he followed it down himself. Then he walked off carrying his coil and playing out the wire until he reached the spot where he had established his "telegraph office." Wherever it was he could hear everything that went over the wire. He could ground it in either direction and could, by using proper precautions, transmit telegrams, as if they were coming from any point along the line, to suit his purpose.[7]

According to this reporter, in order for the new wiretapper to be successful he had to be a thorough electrician as well as a Morse operator of the very first class. A poor-quality sending operator would be picked out immediately upon his trying to transmit a message and a receiver of the same quality could not interpret the dots and dashes as they flew from the key of an expert sender. The single-wire system of telegraphy was the simplest one, and the easiest one to tap. Then there were duplex systems of different patterns and all requiring receiving sets of a like nature to make them work. Next in difficulty came the quadruplex line, which meant there had to be four men at each end of a wire with all of them working at the same time. But even that system, although complicated, could be tapped under certain circumstances, observed the journalist. A potential tapper needed the mileage resistance of that particular wire, the exact mileage point where the tapper planned to operate, an accurate amount of resistance to be used on each side of the intruding telegraph instrument, the corresponding amount of battery necessary to have the same strength, and eight competent men to work it successfully. When telegraph wires were placed increasingly underground officials of those telegraph companies congratulated themselves that their quotations and news were at least safe within city limits (wires going out-of-town remaining above ground). However, tappers soon discovered that it was easier and safer to tap underground cables than to tamper with overhead wires.[8]

This article then went on to mention an earlier instance of wiretapping. It was not often that one telegraph company would deliberately tap the wires of a rival concern in order to steal news, said the reporter, but it was said to have been done in Washington, D.C. It happened during the Baltimore and Ohio Telegraph Company's war with Western Union Telegraph Company some years earlier, probably in the late 1880s. The former company was trying to gain a foothold in the industry but was barred from the stock exchanges, racetracks and baseball parks. They were all controlled by Western Union. The stock exchanges allowed Western Union exclusive rights to distribute their quotations but forbade it to furnish any such quotations to bucket shops. Consequently that class of customer had to turn to the Baltimore and Ohio firm for relief. Whenever a bucket shop was known to be in receipt of market quotations, strenuous efforts were made to discover the source of supply and investigators were hired to unearth the rogue wires and destroy them. It took several weeks for Western Union officials to find out the method by which one bucket shop was getting its quotations. They could see an operator sitting at his table with a pair of sounders, but where he obtained his material was a mystery. That mystery was never revealed to officials of Western Union until the firm bought the Baltimore and Ohio company. A private wire running into the quotation room of a legitimate stock broker on F Street in Washington had been tapped. The tap wires were connected with other wires running north until they terminated in an office on G Street, where a sounder was inserted. From the telephone company a private wire was rented running from the office on G Street to the bucket shop near the legitimate stock broker's premises on F Street, where the operator, by affixing the telephone sounders in his ears, was able to hear every click of the sounder, which was placed directly in front of the transmitter in the office on G Street. As fast as quotations were received he jotted them down and passed the slip to a boy who chalked them up on a blackboard. The office on G Street was kept strictly private with no one entering the premises except those involved in the scheme.[9]

In the middle of October 1895 a news report was flashed all over the country that United States President Grover Cleveland had been assassinated in his summer home at Grey Gables. Within a few minutes thousands of telegrams of inquiry went out in all directions. Most newspapers soon discovered the story was a lie but some were caught and printed the matter as "rumors." Later on the *St. Louis Republic* acknowledged having faked the story. According to that paper the whole thing was a "trap." They thought someone was stealing their news by tapping their wire to New York City and they had therefore fixed up that sensational fake story and sent it out over the wire, hoping the supposed thief would steal it and print it in his paper. The trap did not work but it did unearth something. The New York City correspondent of the *Republic* was a man named Hudgins. He was also a telegraph operator and he said he composed the assassination story in his own head and sent it out over the paper's wire. Hudgins said he had reason to believe someone was tapping and stealing from that wire. He didn't know where that suspected theft was taking place but was under the impression it was at Springfield, Illinois. That wire ran through various intermediate Postal Telegraph Company offices to its main office in Chicago where, explained Hudgins, it was connected by mechanical repeaters with the wire from Chicago to St. Louis, thus making it practically an unbroken circuit all the way through. Hundreds of other wires in the Western Union and Postal offices were connected similarly for private individual service from New York to the Western newspapers. Hudgins based his suspicions on the fact that someone along the line occasionally broke in to ask for the last word or last sentence to be repeated. The sender would do so and the receiving operator was supposed to respond with a special sentence: "Why in thunder are you repeating." That response was a code and stamped the request as legitimate. If no special sentence was used or requested then it was suspect. It was also reported that older, more experienced operators thought Hudgins's story was "silly." The whole thing could be explained by something like a new operator on the job, someone needing practice to get up to speed, and so on. But, it was said, one thing you could rely on was the fact that a real wiretapper never broke in on a message, asked for a repeat, and so forth. The paper's trap was set for 2:30 a.m. just when the morning newspapers were ready to go to press. Several bulletins were sent out: (1) the President had been murdered; (2) the murderer had escaped; and (3) the assassin had been found. Those bulletins were sent out a few minutes apart. The trap caught no wiretapper but it did catch a telegraph operator in the Postal office at Chicago, who may have been honest up to that point. He gave a tip on the story to the *Chicago Tribune* and *Chicago Times-Herald* and he sent the tip over their wires which ran to the Postal office. They queried the *Republic*. St. Louis thought that if the biggest paper in Chicago had the news they must have stolen it, and had been stealing all along.[10]

Frank A. Graham, telegraph lineman, was arrested on October 9, 1896, by the New York Police Department on a bench warrant issued by Magistrate Mott charging Graham with tapping a telegraph wire. It was alleged that Graham tapped a wire between the offices of the New York News Bureau and those of the Stock Quotation Telegraph Company. The offence was a misdemeanor under Section 642 of the New York State Penal Code. When he was arraigned before Mott, Graham pled not guilty and was remanded in custody in default of a $1,000 bail. The tapped wire was one used for the transmission of news between the two organizations and among Boston, Philadelphia, Chicago and other centers. Practically all of the telegraph dispatches received by the New York News Bureau passed over that wire. Graham made a connection with it on the roof of a nearby building and ran it down to a basement room. From the repeater the wire was traced into the office of the Printing-Telegraph News Company, which was also a financial news agency. The manager of New York News Bureau had known for some time that his telegraphic news was being stolen and used in the street before his firm could distribute it to its customers. As a result of those suspicions an investigation was launched and that led to the discovery of Graham and his tapping work.[11] The *Wall Street Journal* gave a brief account of the above case and added, "It has been the experience of the New York News Bureau—and our own as well—that messages have reached the Printing News Co. quite inexplicably."[12]

On the afternoon of November 20, 1896, the Western Union Telegraph Company secured a restraining order in Chicago from Federal Judge Grosscup against the Independent Telegraph Company, which, it alleged, had been tapping the wires of Western Union and other telegraph companies. The manager of the Independent was Oscar M. Stone, arrested one year earlier on the same charge. Stone's firm sold market reports to a large number of bucket shops. Associated with Stone were James W.

Turner, Joseph Moffat, George B. Spangler, H. G. McGill, J. L. Stone, and George H. Stone. According to the order, Stone, Moffat and Turner were the active managers of the Independent. They were charged with selling news and information that had been stolen from the Western Union and Postal Telegraph companies by tapping their wires.[13]

It took almost 18 months, but in Chicago on May 24, 1898, Oscar M. Stone began to serve a six month prison sentence for wiretapping. A reporter observed, "The prisoner is believed to have been the most successful wire tapper in the country. Heretofore he had entirely escaped punishment."[14]

An exposure of the San Francisco telephone company by a reporter from that city's *Examiner* newspaper made a big splash when it was published on January 1, 1899. A subhead of the article declared; "Telephone company has a gigantic system of Espionage," while other subheads stated, "Organized betrayal of its customers carried on for pay," "every conversation at the disposal of the talker's business or social rivals," and "the most confidential messages leaked directly into the office of the *Examiner*." As a lead-in, the reporter observed that the law safeguarded the inviolability of letters sent through the mail—they could not be opened except where the court allowed. "The law, in a long line of statutes and decisions has thrown protection around communication by telegraph. The rights of sender and receiver are defined and trespass upon them is at the peril of the trespasser," he added. He pointed out that the tapping of telegraph messages to secure illicit possession of the messages they carried was made a felony and was punished. The interests of society and civilization were then so dependent upon modern methods of inter-communication—the mail, the telegraph and the telephone—"that any violation of the trust assumed by either method is an offense, not only in itself, but by supplying facility for other offenses, makes the one guilty of it a possible accessory to a wide range and great variety of crimes." He thought there was even more need for the law to stand guard over the telephone wires. The possibility of a leak carried disquiet and consternation into every home, public office and business where the use of that instrument had become a necessity. "A telephone company is an important trustee of the whole community patronizing it, and a violation of that trust is an offense of such aggravated criminality as to require the most drastic reprisals."[15]

A different San Francisco newspaper, the *Call*, published the piece and went on to "demonstrate" that administration of the telephone company in San Francisco was corrupt to the core and its service "a gigantic spy system and organized betrayal of its customers for the benefit of their social, political, amatory, official or business enemies or rivals, who are willing to pay the price of the treachery and violation of privacy." The first offense that attracted the attention of the *Call* was committed by an employee of the phone company in Sausalito. A *Call* reporter had sent in exclusive information on an attempted murder "and immediately the telephone operator repeated it to the *Examiner* correspondent." Upon an investigation the phone operator, it was reported, admitted that offense and was discharged. That incident caused a full investigation into the methods of the Pacific Telephone and Telegraph Company and its subordinate corporation, the Sunset Telephone and Telegraph Company. The former handled the local business of San Francisco while the latter looked after all out-of-town business. Several sources of leaks were discovered but they were found not to be part of the main system. In following them up, however, it was discovered that there was a regular system of leakage, as extensive as anyone wished to have that was willing to "put up." Finally that system of leakage was found in the *Examiner* office. An employee of that paper devised it and then perfected it. His work was to secure control of the operators "and convert the switchboard into an adjunct of the *Examiner* office." According to the story, that man's newspaper work consisted in sitting with a telephone to his ear and listening to messages that were leaked on to his wire by simply bringing his "plug" into contact with the wire that he wished to tap. In that way news reports to the other papers were stolen; private conversations between business men; politicians, public officers, men and women, husbands and wives had been taken off the wire, written down and kept in the *Examiner* office for use "in the kind of journalism practiced there." A prominent California State official had a conversation over the Sunset wire with the warden of San Quentin prison relative to a prisoner named Durant. Each official was alone at his end of the wire but the operator at the switchboard leaked every word onto the *Examiner* wire and the conversation appeared in the newspaper the next day. When Durant died, the undertaker in charge of arrangements received a Sunset message from the Pasadena crematorium that the body would be received there and cremated. Within five minutes the *Examiner* reporter was on the phone asking when the body would be sent to Pasadena and when the undertaker tried to evade the questions the reporter said; "You have just had a telephone message from Pasadena."[16]

In another example, the *Call* telephoned a man in Oakland for his photograph, expecting to run it in connection with a contemplated public distinction that might be awarded to that man. A few minutes later an *Examiner* reporter appeared at that man's house and asked why the *Call* wanted his photograph. As well, it was reported that the city editor of the *Examiner*, "under whom the system of leakage was devised," had a grudge against John P. Young of the rival *Chronicle* and reviewed every telephone message to and from Young and "sought such use of the information so obtained as would be harmful to Mr. Young." When the ferry *T. C. Walker* blew up, the *Call* issued an extra edition. The *Examiner* wanted to know what the *Call* would do and the tapper "was at once plugged onto the *Call*'s wire at the switchboard, and heard the orders issued for an extra." It was also found that the instructions given to telephone operators who handled the out-of-town business over the Sunset wires were to take down verbatim every conversation held over those wires and that every conversation was transcribed, indexed and preserved in the telephone office, presumably for possible future use. "It is alarming that any and every message is subject to leakage and that there is no privacy nor confidence possible," said the reporter and that the situation was the result not of accident but "of a regular system devised and hourly carried out by the employees of the company and under the noses of its officers." In conclusion the reporter mused; "How do Mayor Phelan, Governor Budd and other prominent officials like to know that they have not had privacy for a single message they have ever sent under the supposed security of these wires?"[17]

On the following day the *San Francisco Call* issued an editorial on the subject. The piece noted that someone had exposed the untrustworthiness of the telephone company, "not a pleasing task, but it became a duty and was performed." The editor continued, "That a corporation whose chief hold upon the public is based upon the understanding that it will keep faith, to betray the secret messages of its patrons is a breach of decency. It passes beyond a simple breach of trust. It is dishonorable, dishonest and despicable." When a message was put on the telephone wires it was always with the idea that it would go from sender to receiver without being diverted to the listening ear of a wiretapper. That a man connected with the *Examiner* should have conceived the plan of breaking into the confidence of people conversing on matters in no way his concern was, declared the editor, "not surprising. In doing all that tapping, the editor believed, the man enjoyed the sanction of the company and the active cooperation of employees. "To him were revealed the communications passing between husband and wife, between business partners, between associates in politics. By the scheme of tapping the wires he was able to know many things he had no right to know, and which a man of principle would have scorned to have acquired." However, the editor did not blame the newspaperman alone but thought the telephone company had to share that blame. For it was obvious to him that messages went awry, that messages were put in writing and placed on file. "Why? That the company has no right, no good purpose in doing this, is clear. It is, in the light of revelations, a grievous and impudent wrong, placing every subscriber at the mercy of a corporation." With respect to the scandal exposed in his own newspaper the day before, the editor declared, "The telephone company is culpable. It has been exposed and its disgrace is open. What have its representatives to say?" He called, further, for the "abuse" to be regulated by statutory enactment because "The patron of a telephone company is entitled to the same protection afforded to the patron of a telegraph company, or of a post office. Such dereliction as has been laid bare should be corrected through the enactment of a law making it a felony."[18]

Most residential customers of phone companies were on party lines in this period. That is, two or more people in different houses had the same phone number. A series of different types of rings—such as two long, one short—alerted each person on the party line whether the incoming call was for him. Of course, someone else on that party line could later quietly pick up his phone and listen to the other person's conversation. An article originally published in a Chicago paper but reprinted in several other newspapers at the end of March 1903 told of a strange occupation—that of telephone tapper. A veteran employee of the telephone company told the reporter, "There aren't ten men in Chicago who know what a telephone tapper is, but there are hundreds of persons who have come to grief through his work." The telephone tapper under discussion was a man who was hired by the telephone company. His business was to tap the wires on party lines and at hotels and such places to see if the telephone was being used by persons who were not careful of what they said. "Often the company receives complaints that telephone users say unprintable things that are unavoidably overheard. The company tries to do away with this sort of patronage. Hence, the tapper," explained a reporter. According to that employee, the tapper had to be a man of patience. It was not unknown for one of those tappers to sit for 20 hours at a stretch waiting for a signal. When a complaint was made that the "wrong kind of talk" was circulating on a party line, the tapper went to one of the houses, generally the complainant's, and tapped the wire. That was done with a specially constructed instrument. It was fastened to the regular phone and then the tapper sat back with the receiver clamped to his ear to await a call. (He was not actually sitting in the home of the complainant but ran his tap wire outside to somewhere else.) "He takes notes on every conversation he hears, and sometimes he must repeat his vigil day after day," explained the article. Recently a complaint had been made by a man about a party line. He said a "very disgusting courtship" was being carried on over the phone and his wife and daughter could not take down the receiver without hearing something they should not hear. In that case the tapper

Illustration of how a San Francisco phone company was "betraying" its customers in 1899 by allowing others to access phone call information. One instance involved a reporter for a city newspaper tapping the line, with the help of the phone company operator. A husband and wife are shown conversing (through that operator) while the reporter is at his distant office, tapped into the line.

went out, tapped the line and waited. All afternoon he got nothing except orders to the butcher, grocer, and so on. Finally in the evening the bell rang three times. The tapper looked at his notes and learned the call was for the home of a well-known family. It was a young man phoning a young woman at the house. The tapper listened "and heard a conversation that I would not repeat." He let the couple finish their conversation and then returned to his company's complaint office. The next day a notice was served on that household to the effect the phone must not be used as it had been in the past. Another complaint involved a man prone to swearing on the phone. This case was solved as in the example above. "Of course, it very often happens that the tapper waits vainly for his parties, but he hears enough of the private affairs of people to fill a dozen such note books as he carries," wrote the employee. He added, "If the people who use telephones knew that they are telling their stories to a tapper as well as to the person at the other end of the line they would be more careful." However, he declared that tappers, like dead men, told no tales. But it was also stated that the tapper "keeps a record of what he learns and in the records are the names of some people who are supposed by their friends to be of the goody-goody sort. It's a peculiar kind of work at any rate, and one of which the public knows nothing."[19]

From the headquarters of the New York Police Department on May 14, 1903, came the announcement that William N. Amory, who had sued President H. H. Vreeland of the Metropolitan Street Railway Company for libel, had told Police Commissioner Greene that detectives, presumably acting for the railroad company, had been shadowing him ever since the beginning of his actions against that company. His wife had also been followed, his mail tampered with, and his telephone wire tapped. Greene said he told Amory he could not interfere in his quarrel with the Metropolitan and that if Amory believed he was being "annoyed intentionally" he should go to a magistrate for a warrant. After leaving the New York Police Department headquarters Amory told reporters; "The telephone people told me my wire had been tapped and while they now have a volt meter on it, I'm not absolutely safe yet. In going to Commissioner Greene, my object was not so much to complain as to get on record this condition of affairs."[20]

A court case involving the placing, or not, of the New York Building-Loan Banking Company into receivership came up in court in August 1903. The firm was said to be the largest saving and loan institution in New York State. David Robinson was its counsel. In court he said that when the state examiners were in charge of the company's offices, the telephone wire was tapped and stenographers made records of a conversation in which State Bank Superintendent Kilburn instructed the examiners to get out their report, even though the accounts did not balance, and not to delay the matter. The state of California wanted to put the institution into receivership while the bank opposed such a move.[21]

When Kilburn took the stand to testify, he said he had not entered into a conspiracy to wreck the company, that there was not the slightest truth in the allegation, especially that part of it supposedly based on information caught by the financial institution when it tapped its own telephone wires while Kilburn was at work in the bank's offices. Continuing, Kilburn said that he had not been ignorant "of the extraordinary proceedings of the company's employees in tapping the wire and listening to what passed between him and his subordinate through the telephone." His deputy, a Mr. Skinner, had warned him of such a likelihood of wiretapping but Kilburn said he told Skinner he had nothing to say to his examiners that he would not care to have the bank hear. Kilburn added, "Before the company hit upon the plan of tapping the wire the State examiners had considerable trouble in getting at the true state of the company's affairs."[22]

In August 1903 in Austin, Texas, the Texas Supreme Court rendered a decision affirming the judgment of the lower court wherein it awarded the Uvalde National Bank of Uvalde, Texas, $1,200 damages against the Western Union Telegraph Company for a loss sustained through wiretapping. The money was obtained from the bank by men named Rief and Fisher, described as "two noted wire tappers," who were then serving sentences in the Missouri penitentiary. The Supreme Court held that Western Union was negligent in the transmission of the message over its line. Rief had played the role of a cattle buyer and the pair had tapped telegraph wires in order to fool the bank with fake messages.[23]

John H. Gibbons, a telegraph operator who had formerly been an employee of Western Union's racing department, began a suit in the Supreme Court of Brooklyn on February 2, 1905, against M. J. McKenna, F. W. Flood, and J. H. ("Panama") Kelly for an alleged breach of contract in the division of the profits obtained by securing racing information from the New Orleans racetrack. The complaint of Gibbons disclosed a seemingly gigantic wiretapping scheme that had been in operation up to the time of the recent purchase of the National News Association by Western Union. The information received by the tappers was transmitted, it was alleged, to the New York office of Western Union's racing department where McKenna, assistant to the head of the department, sent the news by private wire to the headquarters of the combine, from which place it was sent to seven pool rooms in Manhattan and Brooklyn. Gibbons said the director of Western Union had no knowledge of what was going on in the racing department. All the messages were sent to McKenna personally and they were handled entirely by him. He said Western Union did not have anything to do with the matter and that therefore there was no violation of the agreement between the company and the city officials made the previous racing season in which Western Union promised not to disseminate racing news to pool rooms in New York City. McKenna was formerly the Western Union correspondent for the New Orleans racetrack. When the racing season began, said Gibbons, Kelly was sent down to New Orleans by the combine to see what he could do in the way of tapping the wire of the National News Company. Kelly succeeded in tapping the wires, thus giving to the combine the information that it cost the National News Company $450 to obtain. Gibbons asserted that the partners failed to divide the profits with him and he announced that he intended to place all the information in his possession into the hands of the Western Union Telegraph Company.[24]

7
Wiretapping Labor

While it was usually the case that labor groups were the targets of wiretapping, sometimes the reverse took place. During a railroad strike in 1877 the strikers held a meeting in Grafton, West Virginia, on July 19 and resolved to send assistance to their comrades at Keyser, where they had learned that small squads of United States troops had been sent to guard a train that had just arrived. The strikers in Grafton were said to be 100 strong and "beyond the control of the civil authorities." The telegraph wires between Martinsburg and Wheeling, claimed the account, "have been tampered with by the strikers, who have among their number some men who are operators. It is known that they have tapped the wires to get information of the plans devised to circumvent them."[1]

In St. Louis, Missouri, on April 5, 1886, Frank McKeighan, a telegraph operator, was arrested and made, it was reported, a confession. He said that recently he had entered into an arrangement with Thomas Furlong, chief of the Gould System secret service, whereby Martin Irons, A. C. Coughlan, and other prominent leaders of the Knights of Labor organization were to be arrested. Jay Gould was a leading railroad developer and speculator of the period and was the archetypical "robber baron." The Gould System secret service was his force of thugs that brought terror to organized labor whenever it could, usually with the aid and blessing of the state. The Knights of Labor (officially the "Noble and Holy Order of the Knights of Labor") was the largest and most important American labor organizations of the 1880s. It involved itself in social and cultural issues as well as with economic issues. According to McKeighan's story, a room was rented on the third floor at 23 Market Street in St. Louis. Going past the window of that room was a private telegraph wire over which H. M. Hoxie, first vice president of the Missouri Pacific Railway, was in almost constant communication with Gould. McKeighan engaged another telegraph operator named Nichols to assist him and the two men soon tapped the wire and had a telegraph instrument at work. The plan was to have Irons, Coughlan, and others in the room intercepting dispatches between Hoxie and Gould, when the police would make a raid on the place and capture all those high-ranking labor leaders. It was, in other words, designed to be a set-up to ensnare labor leaders. The signal for the raid was to be a lamp placed in one of the windows of the house containing the rented room. When a light appeared in a second-floor window the police crashed into the room—but found only a solitary woman working away on a sewing machine. When it dawned on the police that the lamp was in a window on the wrong floor they rushed up the stairs but found only McKeighan there. He was arrested. A railroad strike was in progress at the time and a sub-headline on one of the articles about this story declared, "Gould's private wire tapped by strikers," despite the fact that no such thing had happened.[2]

As the story continued to unfold it got more and more confusing. An article two days after the above incident stated that more facts were coming to light that "reveal a deliberate plot to entrap the leading men of the strike and as was suspected connect certain Western Union officials with the scheme." McKeighan, described therein as "the tool of the conspirators," had by this time retracted all of his former statements with respect to this incident; in fact, he was then saying he never even made those earlier statements and claimed he had acted on his own responsibility. As well, he denied he was acting either for the Knights of Labor or for Detective Furlong; that he had never been introduced to Irons or Coughlan and that he had never visited them or made a proposition to them to tap the wires. Yet he admitted after the raid on the house on Market Street that he had met Furlong and had a talk with him. In this article it was reported that McKeighan was viewed by his fellow operators in Western Union as an agitator and was prominent in effecting an organization of the Brotherhood of Telegraphers. "The revelations of the past few days makes it clear to members of that order why certain operators have been quietly dismissed without cause," remarked a reporter. "A prominent leader in the brotherhood said that they had long suspected McKeighan's fidelity and had been watching his actions closely for some time." Thus they were not surprised by his arrest.[3]

A different story was presented by Assistant Circuit Attorney McDonald who said that the first information regarding the plan to tap the wires was communicated by Western Union officials to the Missouri Pacific officials and that the latter directed the raid. The information from Western Union, said McDonald, "was that one of the company's employees had been approached by strikers who were willing to pay him money to have him tap the wires, in order that they might know what Hoxie and Gould said to each other. They made a great many applications to him. He had at last accepted, and conveyed the information to his employers."[4]

On the evening of April 16 in St. Louis, four bench warrants were issued by Judge Van Wagoner of the criminal court for the arrest of Martin Irons, A. C. Coughlan, George M. Jackson, and S. W. Nichols, all indicted for tampering with telegraph wires. All four were members of the Knights of Labor, with Irons and Coughlan being high-ranking leaders of the union. Those indictments were the result of the grand jury's investigation of the tapping case, said a newsman, "in which the city police showed that the railroad company had attempted to lure the Knights into the commission of a felony." Frank P. McKeighan was described as the telegraph operator "who is alleged to have been instructed by the company to induce the Knights to tap the wire." This version of the incident began by stating that McKeighan, a telegraph operator employed by Western Union and a member of the Knights, was friends with L. L. Richmond, a prominent Knight and a wholesale dealer in sewing machines. Around March 20, 1886, McKeighan dropped into Richmond's store and was there introduced to E. M. Jackson. Jackson suggested McKeighan accompany him to a place where a Knights committee was in session. At that gathering McKeighan was introduced to Irons and Coughlan. Learning that McKeighan was a telegraph operator the subject of dispatches came up and the two labor leaders asked if it was not possible to tap the wires connecting the Western Union office with the Missouri Pacific headquarters. Irons said he had already tapped a wire at Marshall, Missouri, and received considerable information. McKeighan replied that it was possible and was instructed to find the wire and the Knights would furnish a man to do the tapping work. Later McKeighan reported back that the wire was in a cable and that an expert was needed for the job to be done properly. Coughlan said he had such a man and introduced McKeighan to S. W. Nichols, a telegraph operator who had come to St. Louis from Nevada, Missouri, by order of the Knights of Labor. As a next step McKeighan reported the scheme to Mr. Brown, the general manager of Western Union in St. Louis, and was instructed to go ahead and play along with the scheme, and report its progress to Western Union. McKeighan next informed Irons that the Missouri Pacific had requested Western Union to make a loop from the main office to a certain drugstore so that Hoxie could receive night messages directly: in his estimation, the loop would offer greater inducements to tap than did the cable. He was instructed to find the wire and did so, tracing it from the Western Union office to the drugstore, which was a few doors down the street from Hoxie's home. Around that time McKeighan was shown a telegram by Irons which he said had been intercepted along the Missouri Pacific line. The operator made a copy of it and gave it to Brown. The manager took that telegram to ex-judge Portis, general solicitor for the Gould System, and asked the co-operation of the railroad in discovering how the dispatch was intercepted. He then told Portis of the attempt that was about to be made on the wire in St. Louis. Portis immediately summoned two private detectives from the Gould System, Furlong and McDonald. After consultations it was decided to go ahead and McKeighan told Irons and Coughlan he had located the wire, as they had asked him. Jackson and Nichols checked out the area and selected the house on Market Street. A room was rented on the third floor by Jackson. Friday night, March 26, was set for the task. Meanwhile Brown had been reporting to Missouri Pacific detectives and it was decided to surround the house and raid it that night, using the lamp in the window as a signal. However, due to technical difficulties, Jackson and McKeighan could not accomplish the task of tapping that night. The job was postponed for a few days. Jackson and Nichols went away, leaving McKeighan alone in the room. Then a coincidence spoiled everything. Mrs. Smith, the tenant on the floor below, lit a lamp in her window, leading to the raid on her room. Then, after the mistake registered the raiders rushed to the third floor but no one was there except for McKeighan. He had a discussion with Furlong and McDonald, who were with the raiders, and decided they would go ahead with the postponed plan and not tell the Knights about the raid. However, city police, who were not in on the plan, arrived at that time and McKeighan was arrested. Nichols then disappeared from sight. Irons and Coughlan agreed to appear in court the following day. The Knights of Labor were indignant at the results of the grand jury who, they said, were influenced by Assistant Circuit Attorney McDonald, acting as attorney for Gould.[5]

A few days after that an editorial appeared on the subject. The editor started by noting that Martin Irons and two other Knights of Labor had been indicted at the request of Jay Gould for conspiracy in tampering with the private telegraph wire into Hoxie's office. It was, he declared, "a job put up between the managers of the railroads and their detectives to inveigle leaders of the Knights of Labor into an attempt to eavesdrop on Jay Gould's wires." He continued that "the conspirators were the ones who pretended to be anxious to vindicate the law ... Gould has control of all the lines, and can pry into everybody's most secret messages whenever he feels disposed. Before he got possession of the Western Union through stock jobbing he freely cut and tapped its wires law or no law." He added,

"In this very city of Omaha not many years back a squad of Jay Gould's telegraph men armed with revolvers, climbed the poles of the Western Union, cut its wires and transferred them into Jay Gould's American Union office." The editor stated that he did not propose to argue that Irons and his colleagues were justified in attempting to capture Hoxie's and Gould's telegrams, "but the way in which this job was put up to create a sensation and terrorize the Knights of Labor is on a par with the treachery and double dealing displayed by the king of monopolists in his conference with Powderly [then head of the Knights]. Conspiracies against life are crimes except when committed by millionaires and stock-jobbing railroad managers."[6]

As of late April 1886, four men were under indictment in the affair: Irons, chairman of the executive committee of District Assembly 101, Knights of Labor; Coughlan, who held the same office in District 83; George M. Jackson, a prominent Greenback politician in St. Louis; and S. M. Nichols, telegraph operator. The charge against them was "tampering with the telegraph wires," a statutory offense punishable by imprisonment in the penitentiary for two years or a fine of $500. Those charges were based almost solely on the very contradictory statements made over time by McKeighan.[7]

Nothing more happened until September 22, 1886, when it was reported that Irons had been brought to St. Louis from Kansas City that day by U.S. Deputy Sheriff Skidmore to stand trial on the outstanding charges.[8] After a number of delays the trial took place and Irons was acquitted at the end of February 1888. The prosecutor then announced he would enter *nolle prosequi* (i.e., drop the charges) in the cases of Coughlan and Jackson. That ended the criminal prosecutions growing from the railroad strike of 1886.[9]

On October 4, 1889, Terence Powderly, head of the Knights of Labor, made a speech in St. Louis in which he denounced Thomas Furlong by "proceeding to demonstrate that Furlong had conspired to entice Martin Irons and others to attempt to tap the wires, so he could make a criminal case against the strikers." Furlong was then an applicant under consideration for the position of chief of the United States Secret Service. He did not get that post.[10]

A brief report was published in July 1894 when a railroad strike was under way in San Francisco. The United States authorities declared that the wires between Sacramento and San Francisco had been tapped and instructions from Washington to federal officials in San Francisco given to the leaders of the strike. Hereafter, said the report, all government messages by telegraph would be in cipher.[11]

A strike at the Bunker Hill and Sullivan mine was under way in April 1899 in Wardner, Idaho. According to a report shots had been fired and the place had suddenly been turned into an armed camp. Non-union miners attempting to enter the mine were met by strikers. Those strikers succeeded in keeping the scabs out for one night but not on the following night when Sheriff Young arrived from Wallace, Idaho. "The strikers have tapped the telephone line between the mill and the mine and are intercepting all messages," remarked a newsman.[12]

In June 1916 it was revealed that the New York Police Department had been systematically tapping the phone lines of labor unions. That information came to light during sessions of the Thompson Legislative Committee. George L. Thompson was a New York State Senator. On June 7 New York City Police Commissioner Arthur Woods made an offer through Frank Moss, counsel for the Thompson Committee, to permit representatives of labor unions to see the records on the tapping of phone lines of labor organizations and to explain why each line was tapped. Moss promised Woods he would send that word out to representatives of labor unions before the next scheduled meeting of the Thompson Committee, a few days later. Peter J. Brady, Secretary of the Allied Printing Trades Council, had made the charge that wires to labor union headquarters had been tapped and declared that he wanted the reasons for the taps to be released in public and from the witness stand (that is, under oath) and not in secret and not under oath as Woods had offered. Brady said he wanted the charges thrashed out before the Thompson Committee and thereafter to ask District Attorney Swann to place the evidence of the tapping before the grand jury for possible action. According to Brady, one of the tapped wires was Stuyvesant 1126, the phone number of the International Ladies' Garment Workers' Union, which had been locked out by the employers. Another tapped line was Stuyvesant 4867, the phone number of the Amalgamated Clothing Workers of America. Brady added that both of those lines were tapped before the disclosure of tapping in a parallel and ongoing private charities inquiry and were still being tapped as he spoke: "A number of other wires of unions are being tapped and I will make public other numbers later." New York City Mayor John Purroy Mitchel said, "I do not know whether the wires mentioned were tapped, but I can say that no wires have been tapped except in search of information regarding the commission of a crime and for the apprehension of criminals. The statement made before the Thompson Committee yesterday [by Brady] was an irresponsible statement made, so far as I know, by an irresponsible man." Senator Thompson expressed doubt that his committee had jurisdiction over the new wiretapping charges but that they would decide within a few days when he held a hearing on the matter with the union men. The American Federation of Labor, the New York State Federation of Labor and several local unions were to be represented in the group of union officials who planned to place the charges before the Thompson Committee.[13]

New York District Attorney Swann, aroused over complaints made to him regarding the tapping of phone lines, announced on June 13 that he would begin an investigation into the practice and prosecute those who had been guilty of listening in for personal reasons. Swann had been told, in addition to the allegations from organized labor, that divorce evidence was obtained through listening in on phone conversations by operatives from private detective agencies. He said he had been considering the advisability of opening an investigation for several days and that he came to a decision to proceed after a visit paid him by Brady, who told him of labor union lines being tapped. Brady told Swann, "It is my suspicion that employers of labor hired private detectives to keep tab on labor organizations and these detectives through connivance with police officials were permitted to tap telephone wires."[14]

In connection with the demand of labor organizations made upon the Thompson Committee for the publication of the 350 persons (not just labor groups) whose phone wires had been tapped by the police and conversations recorded, it developed that the list included several "eminently respectable women." That statement came from Frank Moss, counsel to the Thompson Committee, in response to that request for a public disclosure from Peter Brady. Said Moss to Brady, "The trouble is that there are a few ladies in the list, women of excellent reputation, and it would put them in a most embarrassing position to publish their names." In other words, there would be no public disclosure of the names of people whose phone lines had been tapped by the New York Police Department. Brady was accompanied by a delegation of about 30 labor union figures. In their formal letter to the Thompson Committee the union men declared that in justice to themselves and their unions the reasons for the tapping of wires must be made publicly known. Labor men also demanded there be an investigation by the Thompson Committee of the relationship between the New York Police Department and private detective agencies. They asked that the Committee "make public the private understanding that permits the Police Department to allow these private detective agencies to tap wires and install dictagraphs and detectaphones, break into offices, smash desks, and copy private correspondence." They continued, "The public ought to know to what extent the so-called private detective agencies and strike breaking associations are supervised by the Police

Department; what reports they do furnish to the Police Department of their activities." The group added that it was high time the public of New York City realized and understood "the ramifications of these moral lepers [who work in detective agencies] so that steps may be taken immediately to curtail the abuses of the powers which have been given them, in conjunction with the Police Department." In its conclusion the letter referred to the "Russianized secret service prevailing in New York City." That letter was presented by Brady to Senator Thompson and was signed by Hugh Frayne (general organizer of the American Federation of Labor), James P. Holland (president of the New York State Federation of Labor), John Sullivan (international officer of the United Brewery Workers), E. J. James (Brotherhood of Painters), Ernest Boehm (Secretary of the Central Federated Union), and Henry Waxman (United Garment Workers).[15]

About ten days later a news story stated that information gained by the police from tapping labor union phone lines had been turned over to employers and to private detective agencies. One specific example given involved a garment worker who was at the time one of thousands on strike. That man had called up his union headquarters to say his family was in want. Within a few hours his former employer heard of this phone message and hurried a basket of food to the striker's home and followed that up by sending his automobile to take the family out for a drive. "In every case of this kind the effort has been made to influence a striker's wife to persuade her husband to go back to work—but the bosses have had very little success at this," remarked a journalist. With respect to Brady's request for public disclosure of the names of unions and people who had been tapped, and the reasons for such action, there was reportedly much wrangling between Thompson and Moss. In the end it was decided to give Brady a copy of the union wires that had been tampered with and to let him and a committee of labor leaders look through the whole list of tapped wires "upon his promise that he would not publish it." Brady was told that the investigation he demanded (about the relationship between the police department and outside agencies with respect to information received from tapping) was not within the scope of the Thompson Committee and that Brady should take the matter up with District Attorney Swann. That article ended with the reporter noting the list of taps that the Thompson Committee had "was furnished by the telephone company and contains only a small percentage of the numbers of telephones that have been tampered with."[16]

New York City Magistrate House, after an investigation into the tapping of phone lines by the police made at the request of Police Commissioner Woods, reported to the Police Commission on July 30, 1916, that the police were well within their rights in listening in. Criminals, he observed, were quick to take advantage of every new invention that could be useful to them and it would be unfair to deprive the police of the same privilege. "In no case did he find that a telephone had been tapped for private, personal or political reasons," said the journalist. House declared, "In each case the facts show that the police were acting from a sense of duty and well within the line of their work." Eavesdropping on phone conversations, he added, was most important in preventing crime. He stated further that "it is not reasonable to deprive the Police Department of the use of the telephone in an effort to detect the criminal, on the theory that the telephone is private, and that the police have no right to invade its assumed privacy, no matter for what purpose it is being used, and it might be said that, after all, the privacy of the telephone is more imaginary than real."[17]

The ruling by Magistrate House prompted an editorial in a New York City newspaper the next day. Smugly the editor noted, "The nonsense and misstatements which in certain quarters decried the activities of the police in pursuing crime over the telephone are effectively answered in the report of Magistrate House. He had investigated all the cases reported and his approval is sweeping. In no cases were wires tapped for private or political reasons. No legal rights were infringed." (No report was given as to how many cases of phone tapping by the police had been "investigated" by House.) In conclusion, the editor thundered, "That is unimpeachable logic. To permit any other policy to prevail would be to yield common sense and justice before a campaign of hysterics and lies."[18]

At the end of August 1919 M. Schwartz of the Dannenberg Detective Agency in Chicago was arrested in that city for tapping the phone line of that city's International Ladies' Garment Workers' Union headquarters building on Milwaukee Avenue. When asked why he had tapped the phone Schwartz replied that "somebody" employed the Dannenberg Agency to find out what officials of that labor union were talking about at their headquarters. The members of Local 100 were then out on strike. Schwartz rented the room adjoining the union hall and, it was said, had also planted a dictograph or bug. (See Chapter 10.) His bond was set at $4,000 and both the labor union and the telephone company had preferred charges against him.[19]

8
Wiretapping: Other Crimes and Personal Use

The following tale was told by Thomas Furlong, chief of the Missouri Pacific Railroad Secret Service (that is, the company's private police force). It was the story of a scheme conceived of and worked by the express messenger and telegraph operator W. W. McCalla, whereby he succeeded in robbing the Union Express Company of Pittsburgh of $4,200. McCalla, said Furlong, was about 27 years old when he pulled his scam in 1877. He worked on the express car of the train and saw that packages went on and off the train at the appropriate places. When his train reached Pittsburgh he took a seat on the safe in the express wagon and rode through town until he got opposite City Hall. He jumped off then, in spite of the wagon driver's protests, and disappeared. He should have stayed with the wagon until the safe it held reached its destination. It dawned on the company on March 13, 1877, that it had been robbed. The police were called in and then Furlong and Joe Cupples (a special detective for the express company) were set to work to find the robber. That day a new messenger had brought the run in. Tired of waiting for him to turn up with the safe key, the wagon driver reported the matter to Superintendent Bingham. Eventually the safe was opened and discovered to be empty. Detectives on the case started with no idea of that messenger's identity, knowing only that he had tapped the wires and sent a fake dispatch. They followed the wires and found the spot, in rural Pennsylvania. There was a watchman in that area but the tapper told him he had come to examine the wire. This was done regularly and raised no objection. He knew which of half a dozen wires to pick, which led detectives to believe he was a telegraph expert. Once he had successfully tapped the wire he sent a message to the agent at Brady's Bend, asking for the name of the express messenger who was coming in on train number two, signing it George Bingham (the superintendent).

The reply came back that it was Thomas Bingham, George's brother. He then sent a message, purportedly originating in Pittsburgh, to Thomas Bingham, stating that J. C. Brooks would meet him at Templeton. The message continued that Bingham was to turn over his run to Brooks and take his receipt for money and valuables. Brooks would take the run into Pittsburgh. Bingham was instructed to return to Parkersburg and report to agent McClellan who was awaiting him there. Templeton was a place where trains converged, a place for meals. That elaborate message came, of course, from tapper McCalla. When detectives interviewed various people with regard to this case they got a pretty good idea of McCalla's description. In time they identified George McCalla and discovered that he resided with his brother Will, who was a part of the scheme. They traced Will to Palmer, Texas, where he had been an agent for a railroad but skipped out with $1,800. Detectives almost caught up to him in Atlanta, where he went by the name of Lowry. Then Furlong traced him to Savannah, Lake City, and Key West, Florida, and from there to Cuba. From there he had set sail for Rio de Janeiro, Brazil. By then it was July 20, 1877. Later George McCalla was arrested as an accomplice in the robbery. He posted bond but died before his case came to trial. Some seven years after the scam, when Furlong related his story, Will McCalla had still eluded arrest. There were no more references in the papers to Will McCalla.[1]

What was described as a "unique" crime was reported from San Antonio, Texas, in August 1900. Bank robbers used the telegraph line to pull off their robbery. Authorities believed they must have been expert telegraphers. They tapped the telegraph line somewhere between San Antonio and Uvalde and telegraphed the national bank there to cash a draft in the name of C. W. Fisher, drawn on Woods and Son of San Antonio. One of the robbers then went to the bank in Uvalde, identified himself as Fisher and asked if a draft had arrived for him. Told that it had he received the cash (amount unstated) and disappeared from the area.[2]

A similar scheme was nipped in the bud in Kansas City, Missouri, in October 1900. It was blocked by J. G. Stearn, cashier of the City National Bank, and J. T. Hurt, president of the Commercial Bank at Lawson, Missouri. Late in the day on October 11, Stearn received a telegram signed "Commercial Bank" ordering $5,500 in currency. It was shipped Friday on the 9:30 a.m. Santa Fe train. Soon after the train left Stearn was phoned by the Commercial Bank and told it had received a telegram from City National ordering it to pay out $5,500 in currency. Stern said his bank had sent no such wire and Hurt said he had not sent the order for $5,500 that had come Thursday. The Santa Fe train was instructed to carry the money past Lawson, which it did. About noon a man walked into the bank at Lawson and asked if it had an order to pay him $5,500. Hunt said "yes" and asked him to make out a draft; the man was signing it when a policeman arrived and made the arrest. A telegraph operator who had tapped the wires and sent both the messages was also arrested.[3]

A robbery of $700 from a Canadian Pacific Railway train 15 miles east of Vancouver, British Columbia, took place on a Saturday night in September 1904. The robbers were thought to have fled across the border and to be in the United States. A telegram had been received from an unknown source directing the telegraph operator at Mission, British Columbia, to have the express agent at that station open the safe on the train. The idea, apparently, was to suggest that the agent at Vancouver was away, or for some other reason could not open the safe when the train arrived in Vancouver, so it needed to be opened at the last stop before Vancouver, which was Mission. The telegram was signed correctly with a "D" for the dispatcher and was believed to have been sent by the robbers who had tapped the wires and used their own telegraph instruments to send the bogus wire. The safe was opened and emptied before the train reached Vancouver.[4]

B. V. Dunham was arrested at Gettysburg, South Dakota on September 21, 1905, charged with being the principal in a wiretapping scheme by which $3,800 was secured from a bank in that city. A telegraph operator who had acted as Dunham's accomplice gave the police the information that led to Dunham's arrest. The prisoner (also known as F. D. Miles), it was said, represented himself as a cattle buyer of Miles City, Montana. He went to a bank in Gettysburg and applied for $3,800 in cash, giving a Chicago bank as a reference and asking that Gettysburg wire them so as to ascertain his financial standing. His accomplice, allegedly, stationed himself several miles from town, tapped the wire and intercepted the Chicago message to prevent it from going through. Four hours later the tapper sent a reply, ostensibly from the Chicago bank, and of such a character that the money was paid to Dunham by the Gettysburg bank. The arrest of Dunham immediately followed. No outcome was reported.[5]

During the period covered by this book the concept of a hacker did not exist, at least in the way we think of a technological hacker today. But there were a couple of them around. It was reported in the spring of 1888 that an enterprising Dartmouth University freshman tapped the wires of the Western Union and connected them with a telegraphy instrument in his room. He sat quietly in his room and listened to what was going on in the world. Eventually the Western Union discovered the tap after investigating why it was that one of its wires at Hanover, New Hampshire had been partly disabled for several weeks and had been giving the company much annoyance. The job had been done so skillfully with a mere thread of copper wire that its detection was difficult.[6]

It was reported in May 1907 that Hugh Smith, a high school student in San Diego and the son of a well-known attorney in that city, had established a wireless station at his home on Florence Hill and was able to take down every message received at the government station at Point Loma. "His performance has given no trouble but it brings up the question of how the government can protect itself against this kind of wire tapping," observed a journalist.[7]

And then there were the divorces assisted by wiretapping. In January 1889 news articles widely heralded a divorce that was granted due to the telephone. A business man "in one of the cities" had his suspicions aroused because he found his household phone was seemingly always engaged by long conversations when he tried to call home. He had the wire tapped and employed stenographers to take down any and all conversations heard verbatim. A divorce was granted to this unidentified man when he petitioned for one.[8]

Just one month later another such divorce made the headlines, this one offering the full identities of all the participants, and all the dirt. One account described it tersely by observing that the "taffy" gathered by suspicious husband Rutherford Trowbridge at New Haven, Connecticut, from a tapped telephone wire proved sufficient evidence to bring about a successful divorce suit. The co-respondent, Clark Jonathan Ingersoll, a clerk of the Superior Court, had fled the area.[9] Reportedly, the scandal had been settled quietly in court on the afternoon of February 20, 1889, when the petition brought by Trowbridge asking for a divorce from his wife, Anne, daughter of John Anderson, came up. After the petitioner had outlined his case—his spouse did not appear—the divorce was granted. "This is the case where a mass of damaging and very spicy evidence was secured by tapping a telephone wire running to the Trowbridge residence," said a reporter. Ingersoll, who had left town very suddenly, was understood to have fled the area to escape arrest on a charge of adultery. The story got as much media attention as it did because the Trowbridges, the Andersons, and the Ingersolls were three of the "most prominent" families in the city of New Haven.[10]

Although the divorce had been granted the story did not die, not right away. Early in March it was reported that Ingersoll had remained out of the state of Connecticut for about two weeks. He had mailed his resignation as clerk of the Superior Court for New Haven to Chief Justice Park. A later report said he had eloped with Mrs. Trowbridge and the pair were living in Washington State. That same report noted that Rutherford

Trowbridge had married Miss May Farrell, daughter of Franklin Farrell, one of the wealthiest men in Connecticut.[11]

At the time of the divorce Trowbridge also instituted a suit against Ingersoll for $25,000 in damages for the alienation of his wife's affections. That suit was dropped for unreported reasons some time later. Not long afterward Trowbridge's lawyer instituted a suit against Ingersoll charging him with the crime of adultery. As of late in 1892 it was reported that Mrs. Ingersoll had procured a divorce from Jonathan Ingersoll and that Ingersoll had married Mrs. Trowbridge. That marriage was reported to have taken place early in October of 1892 in New York and the couple had gone on to Omaha where they planned to reside.[12]

Learning of his wife's clandestine affair with his chauffeur by tapping a telephone line and hearing them talk, Louis Segal, a wealthy vaudeville promoter in New Haven, Connecticut, in November 1911 urged her to marry the chauffeur, promising to be best man at the wedding and to provide them with a home in which to live. He was then suing his wife for a divorce. Segal was "several years older" than his wife who was 28. The chauffeur, Louis Mann, was 21, or at least that's how the story went.[13]

Peepholes, periscopes, tapped telephone lines, and private detectives were all used by Walker F. Kupfer in 1920, in getting the evidence on which he based his suit for divorce against Ruth Baker Kupfer. A divorce action by Mrs. Kupfer was also pending. She charged her husband with infidelity.[14]

9

Wiretapping: Politics, Laws and Police

One of the earliest uses of wiretapping for political purposes took place in 1868. At that time United States Representative Benjamin Butler was selected as one of the managers of the impeachment of United States President Andrew Johnson before the U.S. Senate. It was not a successful impeachment. As a part of that process a man named Wooley of Cincinnati was brought to public notice by refusing to answer questions relating to his private business. He was charged with contempt of the House and arrested. Eventually he did answer the questions put to him and was released. Alleged frauds were connected with the impeachment and, stated a news story, "There are well grounded suspicions that the conspirators are advised of all telegraphic communications from this point, either through spies in the offices or by tapping the wires, for Butler openly boasted yesterday of having possession of two private messages from Wooley to a friend in New York."[1]

The Brooks-Baxter War (or Affair) was an armed conflict in Little Rock, Arkansas, in 1874 between factions of the Republican Party over the disputed 1872 election for governor of Arkansas. The 1872 gubernatorial election saw a narrow victory for Elisha Baxter over Joseph Brooks. In 1874 Brooks was declared governor by a judge who averred the results of the election had been fraudulent. Brooks took control of the government by force but Baxter refused to resign. Each side was supported by a militia of several hundred men and several bloody battles ensued. Finally, U.S. President Ulysses S. Grant reluctantly intervened and supported Governor Baxter, bringing the affair to an end. At one point in the affair, on April 18, 1874, it was reported that "Baxter still holds the telegraph office but Brooks has tapped the wires in front of the State House, and has an operator in that building."[2]

James A. Garfield served as United States president from March 4, 1881, until September 19, 1881. On May 16, 1881, both of New York's U.S. senators, Roscoe Conkling and Thomas Platt, resigned their seats. Conkling was the instigator of that action, and got Platt to go along with him. It was a political move by the pair that had been festering since Garfield won the 1880 Republican presidential nomination. According to an article published a few days after those resignations, the general belief was that Garfield knew of the resignations by the pair before they were communicated to the Senate. That was said to be because the telegraph operator at the White House tapped the wire and used dispatches not intended for any inmate of that institution. That called forth an explanation by the Western Union Telegraph Company, whose officials said they had no wire at the White House, but that the government had a wire there that made a circuit with the Capital and the various government departments: "The operators at the White House can hear and understand all telegrams sent from the Capitol or any Department." The writer of this news article wondered if that was the usual practice. "If so, then the President may station a man at the telegraphic keyhole and have all that may be privately communicated by officers of the Government at the Capitol or the Departments to their friends or correspondents laid before him for inspection."[3]

A couple of weeks later a newspaper editor declared that Garfield knew of the resignations even before they were received by the rightful first recipient, Governor Alonzo Cornell of New York State. "Can it be possible that the President exercises such presumption? Can he tap the wires at will and thus know the secrets that law makes inviolate?" the editor wondered. "The charge is a grave one, and its truthfulness ought to be verified. The Czar of Russia may do such things in self-defense, but if the President of the United States do so, it is simply impertinent and lawless curiosity."[4]

As a result of the Trowbridge-Ingersoll divorce case some two years earlier, Connecticut passed a law against tapping in the fall of 1889. When Trowbridge set out to get evidence against his wife it was secured by tapping the phone line and eavesdropping on conversations between his home and the office of the co-respondent in the case, Jonathan Ingersoll. Most of the evidence of infidelity was obtained in that manner but, said a reporter, "it created so much comment that the Legislature of 1889 passed a law forbidding wire tapping."[5]

In March 1893 at the Ohio Legislature the House passed a bill that extended to electric light wires the provisions of the wire-tapping law, so as to make it an offense to tap those wires, in other words, to steal electricity.[6]

In Utah the Legislature passed a bill designed to punish anyone who would tap a telephone or telegraph wire. House Bill 127 was passed on the recommendation of the committee on private corporations. It was said to be a copy of the New York State law and was called "A Bill for an Act Relating to and for Telegraph and Telephone Companies." It read; "That whoever shall willfully and maliciously cut, break, tap or make any connection with, or read, or copy, by the use of telegraph or telephone instruments or otherwise, in any unauthorized manner, any message, either social or business, sporting, commercial or other news reports, from any telegraph or telephone line, wire or cable, so unlawfully cut or tapped in this territory; or make unauthorized use by the same, or who shall willfully and maliciously prevent, obstruct or delay by any means or contrivance whatsoever, the sending, conveyance or delivery, in this territory of any unauthorized communication, sporting, commercial or other news reports, by or through any telegraph or telephone line, cable or wire, under the control of any telegraph or telephone doing business in this territory; or who shall willfully and maliciously aid, agree with, employ or conspire with, any other person or persons to do any of the aforesaid unlawful acts, which said employment conspiracies shall be followed by any overt act to effect the object thereof by one or more parties to the conspiracy shall be deemed guilty of a felony, and shall be punished by a fine of not less than $50 nor more than $500, or by imprisonment in the penitentiary for a period of not more than five years; or by fine and imprisonment within the limits herein before specified, at the discretion of the court. Prosecutions under this act shall by indicted in any court having criminal jurisdiction."[7]

The legislative committee investigating Ohio Representative John C. Otis on charges of bribery in the recent senatorial election at Columbus held a session in Cincinnati on January 21, 1898, and planned another session in that city on the next day. All of the members of the committee were reported to have been present. It was charged in the resolutions adopted separately by both branches of the Ohio Legislature that H. H. Boyce of New York came to the Gibson House hotel in Cincinnati a few days before the balloting for U.S. senator and made a proposition of bribery to Representative Otis. The members of the committee meeting in Cincinnati spent some time that day examining the telephone used by Boyce and the rooms occupied by Boyce while he stayed at the Gibson House. It was charged by members of the committee that Boyce fled from Cincinnati on Monday, January 10, and that he could not then be located. Before the committee assembled it was given a telephone exhibition. A Gibson House representative placed a call from its office to the Great Southern Hotel at Columbus (headquarters of the anti–Hanna men during the election) and conducted a conversation. Meanwhile, political operative Jerry Bliss and his stenographer were at another telephone downstairs taking down all that was said. "This was done to show how all of the conversation of Boyce with certain parties in Columbus were taken down while Boyce was here and the senatorial contest was going on at the State Capital." It was alleged that the conversations were all taken down without any tapping of wires as the different telephones in the Gibson House were all on the same circuit. Horace B. Dunbar, president and manager of the Gibson House, testified and confirmed that Boyce had occupied Room 226 from January 7 to January 10. Dunbar said his suspicions were aroused and he set the clerks to watch Boyce. One of the day clerks, Allen O. Myers, Jr., arranged for stenographic reports of conversations Boyce had over the phone during the day. Russell Pryor, the night clerk, recorded night calls. Boyce was given the key to the private office on the second floor whenever he wanted it; the arrangements downstairs were to record all his calls. The stenographer, who had a desk in that office, was always called away whenever Boyce called anyone up. In one of their conversations Major Dick asked Boyce about giving Otis money and what would happen if Otis betrayed the situation. Boyce replied, "I would kill him." Dunbar said he was a Democrat, intimately acquainted with Bliss, Charles Kurtz, and others who had been recently engaged in the opposition at Columbus to Senator Hanna. That was how those opposed to Hanna got their alleged bribery evidence.[8]

The committee proceeded from Columbus to the Cincinnati home base of Representative Otis. It was charged in the resolutions adopted by both branches of the Ohio Legislature that Henry H. Boyce of New York had come to Cincinnati a few days before the balloting for U.S. senator began in Columbus and made a proposition of bribery to Representative Otis. After paying his bill on Monday January 10, and before departing, Boyce told Dunbar that if he wanted to win some money, to bet on Hanna. Several people on both sides of the senatorial contest were guests at the Gibson House from January 7 to January 10.[9]

Some New York Police Department officers were sent out on September 3, 1908, to round up policy sellers. Policy was a form of gambling wherein the bettor picked a series of numbers, trying to match ones randomly drawn a day or so later. A number of these gamblers were arrested in various separate incidents. In one case, Detective Wing thought he sensed a policy shop in the rear of a money exchange office in the Bowery. He went to the rear of the place next door and tapped a telephone wire from the cable box on the back fence. After listening in for a time he raided the place and arrested a man who gave a fictitious name and said he was a broker. He was held in lieu of $1,000 bail.[10]

On the evening of April 30, 1909, New York Police Department officers raided what they believed to be the headquarters of all the pool rooms in greater New York City. In addition to supplying all the pool rooms in Manhattan, Brooklyn, and the Bronx, information on all races was also sent to distant towns in New Jersey. All kinds of telegraphic and telephonic instruments were confiscated. For the previous three to four weeks the men who controlled the pool rooms in New York City had been frustrated to discover how it was that as soon as a new distribution bureau was opened, the police got wind of the location. Suspicion came to rest on some of those in the pool rooms' own ranks and it was suspected there might have been more than one traitor in the ranks. But the information came not from insiders but resulted from the work of detectives Bernard J. Devanney and Thomas J. Conlon, both of whom were described as "expert wire operators." For that reason they had been selected to find the headquarters of the pool room interests. "For many days these two men have been systematically tapping telegraph and telephone wires," wrote a journalist. They went to the roofs of buildings and tapped wires. They scraped a wire with a knife until the insulation was cut off and then applied their pocket telephone. "This is nothing more than common wiretapping and is illegal under ordinary circumstances, but Devanney and Conlon did not hesitate at that," noted the reporter. After several days of fruitless searching they found the correct wire. For several days thereafter they listened in on that tapped wire, and then came the raid.[11]

In court in Buffalo, New York, on October 17, 1909, James B. Waddell of Ohio was sentenced to not less than one year or more than two years of hard labor in prison. Waddell was one of three alleged wiretappers captured in a house one night where they were found with telegraphic instruments connected to a wire that was tapping a line belonging to the Federal Telegraph and Telephone Company. It was a heavy sentence compared to most handed out to tappers.[12]

The trial of George King, a private detective, charged with the extortion of $5,000 from Mrs. Fay Tyson, wife of Robert H. Tyson, began in court on January 4, 1912. R. Buckner, Assistant District Attorney, read a stenographic record of a telephone conversation between King and Mrs. Tyson regarding a demand for money by King in return for his smothering evidence against Mrs. Tyson while she was at Atlantic City with Henry White, her lawyer. King said he had evidence about Mrs. Tyson having an affair with White but he would keep it out of court if he got the money. Robert Tyson had filed a petition for divorce against his wife. The money was paid over to a friend who passed it on to King in the lobby of a New York City commercial building. Just as King was opening the envelope full of money New York Police Department Officer Fitzsimmons who, said a reporter, "had tapped the telephone wire" and listened to King's conversation with Mrs. Tyson, stepped up and arrested him. No outcome on the case was reported.[13]

Arrested in Seattle on November 15, 1912, was Arne A. Nordskog, a private detective. He was detained on a charge of tapping a telephone wire that led to the Burns Detective Agency in a Seattle office building. Evidence against Nordskog was gathered by the Burns agency. Nordskog, it appeared, was employed previously by the Burns agency during the investigation of Federal District Judge Cornelius Hanford. He was a key witness against Federal Judge Hanford in the impeachment investigation that took place in the summer of 1912. Nordskog testified that he saw the judge drunk, consorting with women, and so forth. Nordskog was soon thereafter discharged by the Burns agency (for reasons not stated) and then reportedly employed by people interested in saving Hanford; charges of attempts to tamper with witnesses in the Hanford case were circulated in that connection. A man named Thayer was the manager of the Burns agency and he explained that he learned his wires were tapped early in July 1912 and that the tapper had been at work for nearly a month at that time. Thayer allowed the tapping to continue until he could get enough evidence. The suggestion was that certain evidence in the Hanford case was secured off the Burns phone line and turned over to Nordskog's new employers, who were never identified. It was alleged that Nordskog rented a hotel room in the Right Hotel, two blocks from the office building that contained the Burns offices. Nordskog, an expert electrician, tapped the wires going to the Burns office and brought the wire to his hotel room and in through the window.[14]

Nordskog was arraigned in court on information "charging injury to a private utility." On February 15, 1913, he was convicted of injuring a public utility. The offense was a misdemeanor, punishable by fine and/or imprisonment. According to the prosecution he had been in the employ of interests that sought to discredit or recall Seattle Mayor Cotterill. On December 1, 1913, the Washington State Supreme Court, holding that the evidence in the trial did not show any telephone messages delayed or intercepted or that the property of the Pacific Telephone Company was injured, reversed the conviction of Nordskog.[15]

William Sulzer (1863–1941) was the 39th governor of New York State and a long-serving congressman before that. He was the only New York State governor to be impeached. Sulzer served as United States representative from 1895 until 1912, resigning from Congress on December 31, 1912, having been elected governor of New York in November 1912 for the term beginning on January 1, 1913. Once he was in office a conflict broke out between Sulzer and Tammany Hall; a faction of the Democratic Party. Sulzer remained loyal to the state Democratic Party. He was impeached in August 1913 and found guilty in October 1913 of filing a false report with the secretary of state. On October 17 Sulzer was removed from office. He was the first person ever to be impeached for acts committed before taking office. According to an article published in October of 1913, dictographs were installed in the "People's House," as Sulzer had renamed the governor's Executive Mansion. His political enemies listened to everything he said and "his telephone was tapped." A letter from Chester C. Platt, secretary to Sulzer, confirmed the discovery of dictographs. Wrote Platt, "They were installed not only in the executive mansion, but in the executive offices in the capitol as well. Besides, the telephone wires were tapped and every telephone message sent out copies." A dictograph expert from New York City was called in and unearthed those planted bugs. Spies were said to be everywhere and even the governor's servants were suspected. Colonel Amory, a long-time friend, said, "To tap the wires the agents of Sulzer foes must have obtained charts of the telephone system leading to the mansion. The dictagraphs may have been installed even before Sulzer took possession of the house." He continued, "In other words while the powers that nominated and supported him were hailing him governor of the Empire State, they were literally 'laying the wires' to remove him should he become dangerous to their interests. I have good reason to believe that detectives were after him as soon as he was nominated." Amory added, "Not only the governor himself, but everyone who visited him was shadowed. It was probably the most complete and astute system of surveillance to which a representative of the people in high office was even subjected."[16]

A huge scandal that involved wiretapping, the New York City Police Department, politicians, and various well-known society figures became public knowledge in the spring of 1916. Senator Alvah W. Burlingame declared at Albany, New York, on April 18 of that year, "The admission by the police that they tapped the telephones of Dr. Potter and Father Farrell is the best argument that I know of in favor of the passage of my bill now before the Senate." His bill made it a misdemeanor to overhear a telephone conversation without the knowledge and consent of the parties to the conversation. It was then in its third reading in the New York Senate and passage of the bill was expected soon. The bill was an act to amend the penal law in relation to overhearing phone conversations. Section 1, Article 70 of the New York State Penal Code was amended by adding at the end a new Section, 722, that read, "Any person who, with intent to overhear a conversation on a private or public telephone line, uses any device without the knowledge and consent of the parties to such conversation, is guilty of a misdemeanor." Senator Burlingame said the admission of the police and phone officials "was an outrage" and it if was allowed to go on unchecked there was nothing to prevent the invasion of the homes of people. New York City Police Commissioner Arthur Woods came to Albany on April 18 to have a conference with New York State Governor Charles Whitman. Some politicians believed that he had urged the governor to veto the Burlingame bill if it eventually reached him. A few days earlier Whitman had denied any knowledge of the wiretapping as an aid to the investigations being carried on by Commissioner Charles Strong into the management and affairs of the New York State Board of Charities.[17]

New York State Governor William Sulzer, in 1913. His phones were tapped and his offices bugged and he was physically tailed. Sulzer was likely one of the most surveiled politicians of his era.

New York City Mayor John Purroy Mitchel told reporters that same night that he himself had consented to eavesdropping by the police on the telephone lines of three witnesses in the charities investigation before Commissioner Strong. According to Deputy Police Commissioner Lord the wiretapping was "absolutely necessary" to confirm a complaint that perjury had been committed and was to be committed in the hearing before Commissioner Strong. Mitchel said the New York City Police Department had acted upon information furnished by Commissioner Kingsbury of the Department of Charities. Mitchel added that Kingsbury first came to him with his suspicions and after talking with the charities commissioner, Mitchel said, he advised him to place the matter before Police Commissioner Woods, who thereupon ordered the eavesdropping with the Mayor's consent. Said Mitchel, "On information that a crime had been committed the police listened to telephone conversations over three telephone wires. No other telephones were interfered with. This the police did under authority of law." Phone lines tapped led into the homes of Dr. D. C. Potter, his son Dean Potter, and the Rev. W. B. Farrell, pastor of the Church of Saints

Peter and Paul, all living in Brooklyn. Police said none of those three men was implicated in perjury allegations but that evidence on a fourth, unnamed person was sought. As to the right of the police to engage in wiretapping Lord said the police, under the statute, had full authority to tap telephone wires whenever the detection of a felony made such a step necessary. In the case at issue, he explained, a "believable" person with a "believable" tale came forward, although that person was unnamed. New York City District Attorney Edward Swann was asked if he thought, in this case, the right to tap had been abused. He replied, "I won't say. I'm glad I had nothing to do with it. I can't conceive of myself doing such a thing." (Note that no crime had taken place that was known of.)[18]

John L. Swayze, general attorney for the New York Telephone Company, told the district attorney that his firm had received three letters from the New York Police Department, each over the signature of Police Commissioner Woods, one on March 1, 1916, one on March 21, and one on March 25, stating that it would be necessary for the "purpose of detecting a crime" to tap the wires of certain persons. The phone company agreed to the request with the tapping done at the firm's central headquarters and with the tap wire run into a police station where a policeman listened to messages as they passed over the tapped wires. According to Frank H. Bethell, vice president of the New York Telephone Company, it had been a practice of the police to resort to wiretapping as a means of detection for 10 years. Lord also admitted that was true. Bethell stated the phone company agreed to eavesdropping of that nature whenever Commissioner Woods requested it over his written signature. Swayze declared he was "astounded" when he learned from the newspapers who the tapped lines belonged to. But he admitted that it would not have made any difference to the phone company if the identity of those persons had been known to them before the wires had been tapped. If the phone company was to be believed, they never knew the names of the people to be tapped, only the phone numbers. (It is unclear whether the physical tapping done within the telephone building was carried out by phone company employees or by the police. They had their own "tapping experts" on the force.) Swayze continued by saying that if the police commissioner wanted to tap the wire of J. P. Morgan or Theodore Roosevelt, the company would not be justified in inquiring into the motives so long as the Commissioner stated in a formal letter that the purpose of the wiretapping was "to detect a crime." Dr. D. C. Potter said he had put the matter in the hands of his attorney but there was no doubt in his mind that the purpose of the wiretapping was merely to find out whether Father Farrell wrote the pamphlet attacking Mayor Mitchel's administration and the Strong Commission investigation into the financial affairs of private charities that received some of their money from New York City.[19]

On April 20 evidence offered by the Charities Department to prove that Charities Commissioner John A. Kingsbury was justified in requesting Woods to tap the telephone wires of the three men was ruled out of the charities investigation when Commissioner Strong declined to permit the introduction of reports of conversations over the phone. On the preceding night he had gone over the transcripts of those conversations, almost 100 in number. He said questions such as wiretapping and its bearing on the case in hand were for another tribunal. He claimed his tribunal had no jurisdiction to entertain such evidence. The Right Rev. Charles E. McDonnell, Bishop of Brooklyn, denounced the eavesdropping on that same day as "about the most outrageous offence on the constitutional rights of the people that has ever been committed here."[20]

Two weeks later it was reported that in an effort to establish whether a crime had been committed in the admitted tapping by certain officials of the phone line of three men during the Charities investigation, James A. Stewart, branch manager of the New York Telephone Company, was called before the Kings County grand jury on May 4. District Attorney Lewis denied emphatically that political influence from any source was being brought to bear on him to "ease up" in the investigation. Strong had first admitted some phone transcripts to his inquiry. He said he did so under the general rule that evidence may be received by a court even though it may consist of papers illegally taken. Since then his attention had been drawn to an exception to that rule "on the notion that it is best to falter in the search for truth when success in the search cannot be attained without jeopardizing the interest of society at large."[21]

Controversy over the wiretapping continued and it was reported on May 15 that a Public Service Commission inquiry into that tapping might or might not take place. Mayor Mitchel contended the committee lacked authority to conduct such an inquiry. It was also reported that some members of that Public Service Commission "fear the subject is charged with dynamite and say it is no legitimate concern of theirs."[22]

If the Public Service Commission would not touch the tapping, then someone else would. New York State Senator George L. Thompson's committee investigating public utilities heard on May 16, 1916, that the New York Police Department had been tapping wires "by wholesale" and the committee would begin on the next day an investigation of the entire subject. An official of the New York Telephone Company told Thompson that in the previous two years 350 phone lines had been officially tapped by the police. According to the information furnished to the Thompson

Committee, telephone wiretapping was begun by the New York Police Department back in 1895 when William L. Strong was mayor. Through the succeeding city administrations the practice had prevailed and grown. Senator Thompson said later that he had seen records of 350 instances in which wires were tapped on police request over two years. In each instance, he said, the police demanded the right to listen in on the wires on the ground that they could prevent or detect crime. Frank Moss, counsel to the Thompson Committee, said the wires of "no prominent persons" had been tapped with "a majority" of the places tapped "being those of dives, saloons, or meeting places of criminals." Moss added, "We were informed that the telephone company recognized no requests from private detective agencies—to tap phone lines. They only allowed wiretapping when the applications came from the police in the regular way." And, added Moss, "we understand ... that the telephone company suspects that surreptitious tapping has taken place but the company's O.K. has only been put on the official requests." Coincident with the Thompson Committee's investigation of the subject the Kings County grand jury was said to be ready to again take up the matter of wiretapping in connection with the charities investigation.[23]

On the afternoon of May 18, 1916, in the meeting room of the Thompson Committee, Mayor Mitchel announced that unless Police Commissioner Woods was at once called by the committee to the witness stand he would issue a public statement and "rip things wide open" on the matter on police telephone wiretapping. Mitchel was deeply concerned over statements made by the counsel for Father Farrell of Brooklyn that Mitchel personally had listened in on conversations that took place over Farrell's phone line. The mayor expressed the opinion that the time had come for the facts to be made public. The Thompson Committee had been discussing the wiretapping matter in executive session for some time prior to the arrival of the mayor. Part of the time Woods had been with the Committee, in camera. At 3:15 p.m. Thompson announced he would not call Woods or Mitchel that day. No sooner had he said that than Mitchel and Woods hurried into the meeting room where Mitchel demanded Woods be called as a witness without delay. If he wasn't, the mayor said, he would make his threatened disclosure. The two then adjoined to a private room with Senator Thompson. Loud noises were heard emanating from that room. Prior to Mitchel's arrival Thompson said he believed he had accomplished all the committee could hope to accomplish in the matter of telephone wiretapping and was in favor of dropping the matter right then. Mitchel did not share that belief, apparently, said a reporter. Woods, while waiting to be called, said his testimony would be injurious to the federal government: "I regret it but the Government is bound to be injured by the expose I make. I have got to tell why the New York Police Department has tapped telephone wires. The Federal authorities are working with our local police constantly. That is the reason." Then he went on to say he wanted the chance to tell the Thompson Committee all about police wiretapping, why it had been done, the results obtained and precautions observed to prevent mistakes; "After I have sketched the general aspect of the case, I shall come to the case of Seymour and Seymour and frankly tell the reasons why the telephone of that firm was listened in on.... The method of using the telephone to detect and prevent crime has been seriously injured by the present publicity."[24]

It was speculated that Woods would say the police had been busy for months preventing the carrying out of plans for smuggling arms into Mexico. Since American troops had gone into Mexico, said Woods, "It has been up to me to help the Federal authorities to keep arms out of that country." Seymour & Seymour were a firm of lawyers. Both of the Seymour brothers, Frederick and John, testified that they represented parties (one of their clients was the du Pont Powder Company) interested in a contract to supply ammunition to the Allies, and that J. P. Morgan & Company, bankers, were behind other parties also interested in the contract. Both Seymours said emphatically there was nothing criminal connected with their clients' business. The order for tapping the phone lines of Seymour & Seymour was made out in the regular form, approved and signed by the police commissioner and declared that the wiretapping was necessary to prevent crime. Captain William M. Offley, chief of the Bureau of Investigation of the U.S. Department of Justice, said, with regard to this case and wiretapping in general; "I have never heard of the Seymour case until I read of it in the newspapers. Of course we have worked with the police in a number of cases but have never asked them to tap wires." Meanwhile the Kings County grand jury continued its consideration of the wiretapping scandal. Witnesses heard by them included Farrell, the Rev. Daniel C. Potter, and Commissioner Strong. Mitchel again denied he had ever personally listened in on Farrell's calls. When asked if he knew anything about the Seymour wiretapping Mayor Mitchel said, "I want to say this. Anybody that has been warned that vital interests of the United States as a government were involved and deliberately jeopardizes these interests, while he may not be guilty under any statute of treason to the United States, is a traitor to his country at heart." William S. Butler, counsel for Farrell, reiterated his contention that Mitchel did listen in personally: "My information is that Mayor Mitchel, who admits that he ordered the espionage on my client, went in person to the wire tapping room on Church

Street, established by the police, and listened to Father Farrell's intimate conversations." District Attorney Swann said he was awaiting the receipt of a formal complaint before taking up with the grand jury the matter of the Seymour wiretapping. That law firm had munitions contracts in which J. P. Morgan & Company was interested. It was also reported that it had become known that Gaston Bullock Means, a cousin to former president Theodore Roosevelt, had his phones tapped both at his home, 1155 Park Avenue, and at the Hotel Manhattan, in an effort to learn business secrets. Reportedly, he had been dealing extensively in war supplies.[25]

On May 17, 1916, New York Police Commissioner Woods defended the use of wiretapping for the prevention of crime in a speech before the Wholesale Dry Goods Association in New York City. He said, without mentioning phones directly, "We are not allowed to work any longer in partnership with thieves, although that method was extremely successful in checking the thief's work. We must use the most modern methods and we must use them to the limit. Some of these methods have been a good deal in the public eye and would be manifestly outrageous if they touched honest citizens." He continued, "But they are necessary with thieves. By their use we have been able to meet a thief when he went out to steal and take him before the theft with enough evidence to obtain his conviction while still preventing the contemplated crime."[26]

One day later an editorial on the subject appeared in a New York City newspaper. It began by declaring, "Permission to tap telephone wires and listen to private conversations ought obviously to be restricted with extraordinary care." However, on the other hand, the editor said; "conclusions that all listening by wire is scandalous and ought to be prohibited are hasty and foolish. For the protection of the community and the furtherance of justice it may often be highly desirable and necessary to get evidence by tapping telephone wires." He continued; "Any argument against this method, if carried to its logical extreme, would condemn a private telephone company for furnishing the police or District Attorney with a record of calls, which, as everybody knows, is now an accepted and highly important means of securing evidence against criminals." It was in the interest of justice and public safety for a detective to go into a restaurant and try to overhear the private talk of crooks and conspirators over a neighboring table, one editor argued, so "why should it be any less so to seek to discover and defeat their plots by tapping the telephone?" In conclusion, he stated, "Safeguard by practice against misuse. But don't let excitement get the better of common sense in judging its necessity." No mention was made in the editorial that no safeguards were then adhered to by the New York Telephone Company nor was any request received from the New York Police Department ever denied, provided the request spoke of "preventing" or "detecting" crime.[27]

According to a newspaper article published on May 21, one high city official and a man in civic life would probably be indicted by the Kings County (Brooklyn) grand jury very soon, as a result of the wiretapping of Farrell's phone. At the same time denials came from Washington, D.C., that the Federal Government was interested in the tapping of the phones of the Manhattan law firm Seymour & Seymour. Corroboration of the latter came from H. Snowden Marshall, United States Attorney for that district, who said the Seymour & Seymour wires "were tapped at the instigation of J. P. Morgan & Co." He also said that the Federal Government did not know about the case until the police came forward with information that interested government officials in Washington. Senator George Thompson, before leaving New York City for upstate New York, said that District Attorney Swann should indict William J. Burns of the William Burns Detective Agency for entering Seymour & Seymour's office and taking papers from one of the desks therein. Thompson declared that the facts showed there was neither a national nor an international aspect to the case, unless J. P. Morgan & Company might be regarded of national or international interest. From "reliable sources" it was reportedly learned that Seymour & Seymour and their associates were "suspected of buying purloined cable messages and letters from Morgan & Company" and would soon ask Swann to take criminal action against the invaders of their office. Neither Mitchel nor Woods, who had all along intimated the tapping of Seymour wires was done in conjunction with the Federal Government, had anything to say on the subject at that time.[28]

Senator Thompson said all the talk about national and international issues in connection with the Seymour tap was nonsense. "There's nothing international about it. It was just done because J. P. Morgan & Co. wanted it done, and for no other reason. There was no Federal request to have it done, contrary to the idea that Mayor Mitchel and Police Commissioner Woods would convey." The Senator added, "And we need not be blinded by any false issues. The police tapped the wire of Seymour & Seymour because J. P. Morgan & Co. wanted it done. And what Mr. Burns did in this case he did because Morgan & Co. wanted it done." With respect to Mitchel, Thompson remarked, "Mayor Mitchel is overexcited when he said that I had been warned by a Federal official that probing into the wire-tapping situation would injure the interests of the government. I never received such a warning, nor did I receive any word from any Federal official or any one purporting to represent a Federal official not to go ahead with the inquiry." As far as Thompson was concerned the facts were plain: someone in Seymour's office ran afoul of Morgan & Co. "[a]nd the police and William J. Burns invaded Seymour & Seymour's office and did what they liked." The senator went on to intimate that his Committee was through with the wiretapping situation and that he expected the federal government, as a result of U.S. Representative George W. Loft's resolution, would take up the inquiry. Frank Moss, counsel to the Thompson Committee, disagreed, saying a few hours later that the Committee was far from finished with the wiretapping situation. Moss said the Committee might shortly take up the statement made by a policeman before the Kings County grand jury "that in the last eight years he himself tapped ten thousand telephone wires" (a widely exaggerated statement, of course). "This wire-tapping as a feature of police work, officially recognized as such by Commissioner Woods, is news to me, although I once was in the Police Department. I have been knocking around on police matters for twenty years and never heard of anything so widespread and general as this practice has become," said Moss. "We all recognize that a policeman must get evidence somehow. But the line of decency must not be crossed. You can get a fellow drunk when he has no lawyer to advise him and get evidence that way. And one can also get evidence by listening to a man conferring with his clergyman. But 'white men' don't do that." Frederick Seymour declined to make any public statement on the subject; he had told Swann on three separate occasions that he did not wish to press charges against Burns.[29]

Existence of a plot whereby secret information regarding the buying of munitions for the French government by J. P. Morgan & Co. was stolen from the Morgan firm and sold to munitions manufacturers was revealed, it was said, through the tapping of phone lines going into the Seymour office at 120 Broadway. This fact was disclosed by the insistence of Mayor Mitchel that all the circumstances surrounding the tapping of the law firm's wires should be laid before the public, in order to justify the New York Police Department's wiretapping activities. Reportedly, the mayor practically "forced" the Thompson Committee to reveal the whole story after Thompson had "shown a disposition to delay." Also asserted was the idea that the Seymour wires were tapped to obtain information regarding the alleged shipment by German agents in America of large supplies of munitions to Mexican bandits, although those supplies had been ostensibly purchased by the Allies.[30]

On May 25 it was reported that Frederick Seymour, under pressure, consented to sign a complaint against Burns for the break-in at his office. There was said to be no evidence that the employer of the Burns Detective Agency knew exactly what step the detective had undertaken to ascertain whether the contents of cablegrams about war orders had been diverted from the offices of Morgan to the offices of Seymour, with that information that being sold onward. A newspaper article speculated that Martin Egan, connected with the Morgan firm, had employed Burns. In signing his complaint Frederick Seymour waived immunity from prosecution. It was also reported that George T. Mortimer, president of the Equitable Building Association (whose Equitable Building contained the Seymour & Seymour offices) aided Burns to investigate the Seymour office. Mortimer introduced Burns to Clarence T. Coley, operating manager of the Equitable Building. Mortimer told Coley to do what he could to help Burns. The detective told Coley he wanted to investigate one of the tenants in the building and asked to meet Coley at 8:00 p.m. that night. Coley let Burns into the Seymour office and left him there. On another occasion Burns asked Coley to show him electrical details of the premises. Coley sent for an electrician named Mr. Kidd. One witness called was Bartlett Smith, electrician, whose father, Gaillard Smith, was president of the American Detectaphone Company, maker of a bug similar to the dictograph. (See next chapter.) Smith told of accompanying William Burns, his son Sherman Burns, and a stenographer to a room in the Equitable Building that was next door to the Seymour office and seeing a detectaphone installed. He repeated his assertion that he saw William Burns open a roll-top desk in the office and take from it papers which a stenographer copied and then returned to the desk. With regard to telephone tapping, Smith said his father early in March was asked by the Burns people "in a roundabout way" if he would tap telephone wires. He also said his father had sold a contrivance for tapping phone wires for $75. When asked if the device had been installed in the Seymour & Seymour office Bartlett said, "I don't know. It would have been a very simple thing to set up." When Smith was asked if he thought there was anything wrong when a man asked him to put a bug in the Seymour office he replied, "I don't think when a man hires me." He added, "There is nothing unusual about installing a detectaphone in an office. I have installed, I suppose, a hundred of them, all at the desire of the occupant. There are a great many detectaphones to-day in lawyers' and brokers' offices, and in private homes where there is a suspicion of disloyalty or theft. I always take it for granted when I install one that it is with the owner's consent." After testifying Smith went over to Frederick Seymour to assure him that his only thought was that Seymour himself wanted the bug installed—not that it was being installed to spy upon the Seymour brothers. Said Smith to Seymour; "Why I have put in detectaphones for a great many brokers and buyers and business men who would not talk to a woman alone in their offices without having a stenographer at the other end of a detectaphone to take down every word that was said. That would prevent blackmail." Frederick Seymour explained that he first heard on April 3, 1916, that a detectaphone was in his office and that his phone lines were tapped and that William J. Burns had done it.[31]

When Chief Magistrate McAdoo resumed his John Doe inquiry to find out what crime had been committed in tapping the phones of Seymour & Seymour one of those who gave testimony was Deputy State Comptroller William Boardman, who stated he was of the opinion that the William J. Burns Agency had committed a violation of law (assuming they had done the tapping) and said his office would conduct a separate inquiry. The Burns agency was licensed by the state and that license could be revoked. Evidence had already been given showing that a key was given to Burns that would admit him to the Seymour offices whose wires were subsequently tapped and in whose room a detectaphone was placed.[32]

The New York City newspaper that had published an editorial on the issue of wiretapping returned with a second one, published some eight days after the first one. This one was much more strongly worded and began by saying that "the disclosure of specific instances of promiscuous tapping of private telephone wires in this city has started in the minds of the telephone-using public doubt and misgiving which ought to be promptly allayed." The editor reiterated that the police department "must no doubt arrange for a certain amount of listening in on private telephone wires by department officials or detectives." Then he wondered how many people in the New York Police Department had unrestricted power to tap phone wires and how authorization for the use of such methods was obtained. "How far is it possible for persons provided with necessary instruments to tap ANY telephone wire without the sanction of either the police authorities or the telephone companies?" he further mused. The editor thought those were questions that arose naturally in the minds of the ordinary phone subscriber "who ought to be protected from the risk of having his private business intercepted by listeners whose purposes may little resemble those of justice. Too much opportunity of this sort could hardly fail to become a dangerous instrument in the hands of detectives and police." In conclusion, he stated that the whole situation called for a plain statement from the police commissioner and from the New York Telephone Company showing exactly what restrictions were placed upon phone tapping and what guarantees the innocent telephone subscriber had against unwarranted eavesdropping. "The public is entitled to such reassurance. It should be forthcoming at once."[33]

This whole ongoing scandal, publicity, and various separate inquiries had all started with the New York City investigation into alleged fraud in the funds private charities received from the city to perform their work. That had led to the revelation of the tapping of the phone lines of three men and quickly moved to the revelations involving the Seymour, Morgan, and Burns companies, with more media attention going to the latter incident than to the former one. In that Thompson Committee investigation of the charities the inquiry revolved around an alleged shady payment of $5,000 paid to Daniel Potter, to get him out of his personal financial problems. That took place in 1907 and came out of the budgets of various charities and was designed to look like something else, an honorarium to Potter for services rendered, rather than as a straight transfer of money to a friend in need. Potter was at the time, and had been until recently, the chief examiner at the Bureau of Charitable Institutions, a part of the New York City Department of Finance. His duty as chief examiner consisted of auditing claims and account of charitable institutions receiving city funds. Some charities conspired to defray Potter's personal expenses, it was alleged. At a session of the Thompson Committee inquiry into the charities held on May 27, 1916, members of the New York Police Department "wire tapping squad" were examined. Sergeant Yunge said the squad had been in operation for four years and he had been at the head of it for two and a half years. He reported regularly to the police commissioner and the latter got a copy of all conversations that were taken over the tapped wires. But neither the Commissioner nor any other superior officer ever visited the wiretapping room. Yunge said only a book with a record of tapped wires, including names, addresses, and so forth, was turned over to Deputy Police Commissioner Frank Lord about two weeks earlier. According to Yunge, he was the sole judge of which conversations were important enough to take down and report to the commissioner. Yunge had been a member of the New York Police Department for 15 years and a detective for ten years. When a tapped phone was calling out or being called, explained Yunge, there was a buzz in the tappers' room and the man at the receiver began writing on his slate if the conversation heard was relevant. Yunge explained that the slate method was faster and that none of the men were shorthand writers. They just took "a skeleton of the conversation when the talking is fast, fill it in on the scrap paper immediately afterward and transcribe it by typewriter when they get the time." The Dean Potter line that was tapped was a party line and Yunge admitted that when they tapped one subscriber who was on a party line the police heard the conversations of the others who were also on the party line "but it went in one ear and out the other," Yunge said as he tried to assure the hearing. At the end of the day Senator Thompson announced his committee would cease its inquiry into wiretapping and return to its usual work on "commitments and obligations."³⁴

Testimony before the McAdoo Committee at the John Doe inquiry (the Seymour case) continued late in May 1916. Martin Egan of J. P. Morgan & Co. made it clear that the Morgan firm and nobody else employed the Burns Agency to investigate the alleged diversion of information about munitions orders from the Morgan office to the Seymour & Seymour office. Reportedly, Burns told Egan that in the Seymour office he came across the name of Gaston Bullock Means, the distant cousin of former president Theodore Roosevelt, who was "generally assumed to be an agent of the German Government." Means had complained recently that his phone had been tapped and that he was being followed. Seymour denied

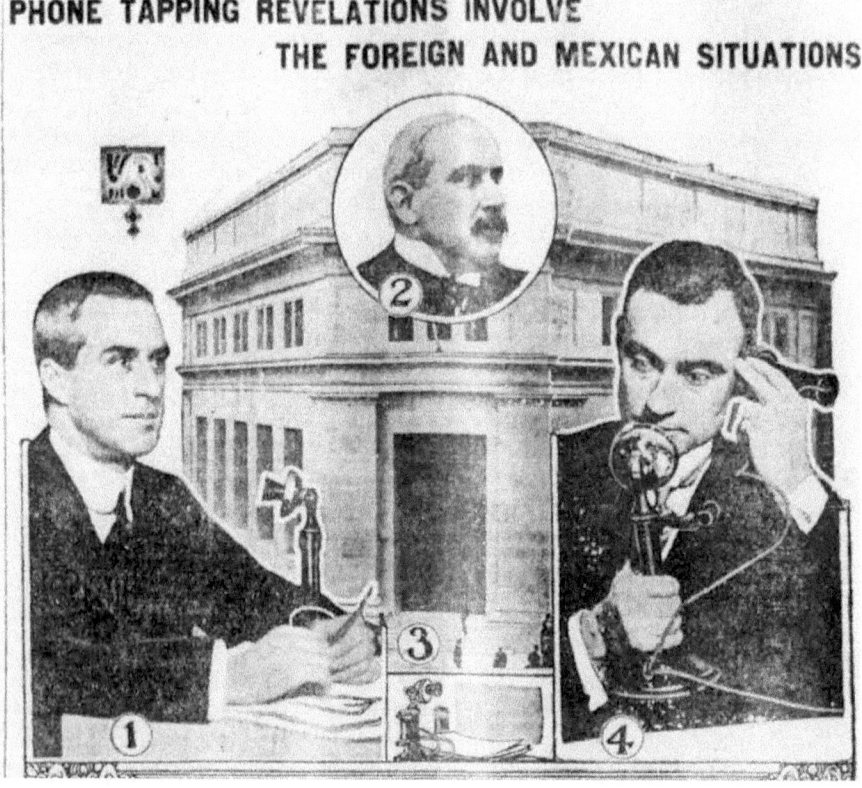

Main players in the 1916 wiretapping scandals. Left to right, Police Commissioner Woods, banker J. P. Morgan, and New York City Mayor Mitchel. The building in the background is the J. P. Morgan & Company bank office.

any and all knowledge of Means. Egan said he also learned from Burns vague information about a shipment of cartridges destined, he thought, for Mexico. Egan said he "assumed" Burns had reported that to the police. Swann said he attached little importance to Egan's testimony and that the day's inquiry had shown no federal side to the Seymour case. It was Deputy Police Commissioner Frank Lord who ordered and supervised the police tapping of the Seymour wires. Testimony showed that tapping "was begun by the police entirely at the instance of J. P. Morgan & Co., acting through Mr. Egan. It was arranged at a luncheon at which were present Commissioner Woods, Deputy Commissioner Lord, Egan, and Julius Spencer Morgan, son of J. P. Morgan and recently made a member of the Morgan firm." That luncheon took place on March 31, 1916, two weeks after Burns had been hired by the Morgan firm to investigate Seymour & Seymour's offices. All of that testimony was given by Lord. After talking with Julius and Egan, Woods told Lord, "You go ahead and handle it." Lord then ordered the tap. Other testimony that day revealed that not only was Seymour's phone tapped but so were the wires of two people in the Morgan office and one person outside who called up the Seymour office. Also, it appeared, a detectaphone was installed in the Morgan office as well as in that of the Seymour brothers. In all, four separate phone lines were tapped and two detectaphones installed. No member of the Morgan firm, said Egan, gave specific instructions to Burns as to his specific activities in the conduct of the investigation. Egan said he did not authorize in advance the entry into the Seymour office by Burns, the opening of a desk therein, and the copying of papers found in the desk, "but of course, I knew they were going in there. I gave them no instructions." Burns, Egan said, had talking about putting a detectaphone in; "He was certain Burns never had tapped the Seymour phone though, of course, he knew the police had." According to Egan, he was employed by Morgan & Co. in a general capacity, with no specific duties. He first heard of the alleged thefts of cable information on March 11 from H. P. Prindle and F. Carrington Weeks, also Morgan employees. They told him that the Anglo-American Cotton Products Company, which was negotiating for the sale of 5,000 tons of cotton linter (involved in the production of propellants used in ammunition) and had been approached by a couple of men from the offices of Seymour & Seymour: William Hills, Jr., first, and then Morton B. Sultzer. Cablegrams were received and decoded by nine men in the Morgan office, which meant a lot of suspects if an internal leak was suspected. On the other hand, the information, if it was stolen, could have been done by outside people tapping the Morgan wires.[35]

Dr. Daniel C. Potter, the principal in the wiretapping controversy of the charities investigation and at one time a Baptist clergyman favored by John D. Rockefeller, died from a heart attack at his office on the afternoon of August 18, 1916. He was found dead in his office at 2:30 p.m. by an employee in the building. Potter was secretary of the Associates of Private Charities (in whose office he died). He was 64 years old. He was still before the courts on charges of perjury and obstructing justice. A decision on those charges had been expected to be rendered within days.[36]

All charges against Potter, and several other co-defendants, in the wiretapping case were thrown out of court by Justice Greenbaum on September 16, 1916. Several grounds were cited—all technical—but the main reason was that Greenbaum ruled the Strong Commission was without the legal power to conduct the kind of investigation that it held.[37]

John A. Kingsbury, Commissioner of Charities, and his attorney, William H. Hotchkiss, went on trial on April 30, 1917, before Justice Jaycox in Brooklyn on indictments charging them with unlawfully tapping the phone wires of the Rev. William R. Farrell and also those of the late Daniel C. Potter and his son Dean Potter. The pair were indicted after a hearing before the Strong Commission, which was investigating the state charities. Justice Greenbaum had already dismissed charges of conspiracy directed against the police officials for tapping those wires.[38]

Delays meant the trial did not take place until May 21, 1917. Farrell and Potter charged that through the New York Police Department and without justification, Kingsbury and Hotchkiss violated their constitutional rights in recording conversations over telephones in their homes. Police Commissioner Woods testified that he had the wires of the three men tapped at the suggestion of either Kingsbury or Hotchkiss, but he could not remember which one, and after Hotchkiss had told him that there was a reason to believe a conspiracy to commit a crime, the nature of which Woods could not then recall. District Attorney Harry Lewis said he would prove that Kingsbury asked Mayor Mitchel to have the wires in question tapped after Kingsbury had learned from William Burns of the existence of the New York Police Department wiretapping squad. Woods also told of the work of the wiretapping squad and that it was housed in a room in a building on Church Street.[39]

On May 24 Kingsbury and Hotchkiss were acquitted in the criminal branch of the Supreme Court, Brooklyn, on the charge they had illegally tapped the phone line of the three men. The jury returned that verdict on the instruction of Justice Kelby who held the evidence failed to show criminal intent on the part of the defendants. Any and all charges pending against the pair were dismissed. And this ended all legal activity in the charities case.[40]

A legal ruling on wiretapping was delivered in June 1916, at the height of the charities and Seymour cases, but independent of them. On June 2 wiretapping for the purpose of detecting crime was held to be justifiable in an opinion handed down by the New York State Appellate Division. The case was an appeal taken by the New York Telephone Company from an order by Justice Daniel F. Cohalan of the Supreme Court directing the company to place a telephone in the saloon of William Restmeyer in Washington Street, upon the payment of the usual charges. Details of the case showed that on January 28, 1916, the telephone company removed the phone from the saloon upon the complaint of the police that the saloon was being used by Restmeyer in conducting a pool room and for the receiving and registering of bets on horse racing. Two detectives had listened in on the talk going on over the wire and later substantiated what they heard by going into the saloon and laying bets. Restmeyer was arrested and later had the charges dropped because the magistrate was not certain that Restmeyer knew that his phone was being used for unlawful purposes. A complaint was also lodged against the phone company. The opinion of the Appellate Division said, "It is certainly not an unlawful or oppressive use of police power to interrupt telephone service by arrangement between the police and the telephone company in cases where the telephone is being used, as in this case, to carry on a criminal business. The telephone company is not required to furnish such service to those who are reasonably sure to use it for an illegal purpose. The telephone company was well within its rights in refusing to furnish its telephone service" in this case.[41]

Also in June, 1916, there was a proposal being made for the erection of a garbage-disposal plant to be built on Staten Island in New York. It was learned that District Attorney Fach of Richmond had been asked to lay before the grand jury the complaint of a number of members of the Vigilance Committee of the Anti-Garbage League that their phones were being tapped. Anning S. Prall and William Woods Mills, two active opponents of the plant, declared that their wires were tapped, as did Borough President Calvin D. Van Name. It was suggested the alleged wiretapping was being done to obtain information as to the proposed strategy to be taken by members of the League.[42]

While Leo Levinsky, a "special agent" of the Tacoma Police Department in Washington State, was phoning to Mrs. Mary Brown, proprietor of the Maze Hotel in Tacoma for a quart of whiskey, which he was to purchase for $3, Detectives Garberg and Modahl tapped the wire and listened in on the conversation. They testified in court, in August 1916, that Mrs. Brown had told Levinsky the package was on its way. A few minutes later

J. Murray, a longshoreman, appeared with the bottle. Mrs. Brown was fined $100 for selling liquor illegally and Murray was given a 30-day sentence in jail.[43]

New York State Senator Alvah W. Burlingame (Brooklyn) offered a bill in Albany in February 1917 on wiretapping that was more drastic than the measure he had introduced in the previous year that had made it a felony to tap telephone wires, and which had been blocked by Mayor Mitchel. The new bill introduced by Burlingame provided that it would be a felony for any one to tap a wire without the written consent of the district attorney of the county where the conversation was to be intercepted. It also stipulated that the interceptors of such telephonic messages had to report to the district attorney the results of the wiretapping within seven days.[44]

Malfeasance in office and the giving of aid and comfort to the enemy were declared to be the charges the Chicago branch of the National Security League was trying to establish in Chicago in September 1917 against Mayor William Thompson. In the eyes of his critics Thompson became guilty of malfeasance when he refused to enforce Illinois Governor Frank Lowden's order to prevent an anti-war meeting the previous Sunday. The aid and comfort charges were said to be rooted in Thompson's "attitude" throughout war operations: he was not enthusiastic enough with respect to the sale of Liberty Bonds. Thompson complained that there was a conspiracy against him: "In furtherance of the conspiracy against me my enemies have recently bored holes in the walls of my apartment, installed dictographs, tapped telephone wires, stationed operators in adjoining rooms and employed spies to hound me."[45]

New York County District Attorney Swann said on October 8, 1917, that a police captain stationed in Brooklyn whose name he refused for the present to make public, had misused his official capacity to tap telephone wires in the Biltmore Hotel in New York on May 17, 1917, for private ends. That police captain had reportedly confessed and a young female telephone operator at the hotel had signed a statement implicating the officer. The purpose of the tap was to obtain evidence against a woman who was stopping at the hotel and whose husband wanted grounds on which to get a divorce. The husband, so the story went, was a friend of the policeman. The woman had a room on the sixth floor of the hotel, while one of the switchboards was on the 22nd floor. The police captain represented himself to the necessary persons at the hotel as being there on official business and a connection was made with the woman's wire in such a manner that he and the husband could hear her conversations. The story of how the police captain used his uniform to further a personal end, it was said, came out at the divorce proceedings when lawyers for the woman sought the source for part of the evidence against her. The captain was to be prosecuted under Section 1423, subdivision 6 of the New York State Penal Code, which made it a misdemeanor for anyone to unlawfully tamper with the wires of communication.[46]

More details were released the next day. The police captain was John L. Falconer. The friend he was helping was Conrad F. Dykeman, a civil engineer then on trial for impersonating a U.S. Secret Service officer. Late on October 9 Police Commissioner Woods announced he was suspending Falconer from duty for tapping wires for other than police department purposes. According to District Attorney Swann, Dykeman, acting for a relative, attempted to get evidence for a divorce against the woman. Swann said that on May 3, 1917, Falconer took Dykeman to the hotel and introduced him to George Williams, house detective, as a Secret Service agent. He stated that there was a woman guest in the hotel who was in communication with an Austrian captain implicated in a plot to blow up the Ashokan Dam in New York State. At the request of Falconer, Williams took the pair and introduced them to Kathryn Haggerty, supervisor of the switchboard exchange in the hotel, and requested her to help them in tapping the phone and intercepting messages for the woman guest. Dykeman then took from under his coat an apparatus for tapping the phone and proceeded to listen in. Later it was discovered that Dykeman was not connected with the Secret Service and that both Falconer and Dykeman had lied. Dykeman was arrested on July 30 by Federal Agent John A. Fenger. He said the divorce matter was a private matter and refused to disclose the name of the woman. On two other occasions Dykeman searched the room of the man whom he referred to as an Austrian officer. At Dykeman's trial Falconer said he did not introduce Dykeman to the hotel detective as a Secret Service man. He did admit, though, that as a police captain he was able to get the permission for the attaching of the tapping instrument on the switchboard. No report was published on the outcome of Dykeman's trial.[47]

Toward the end of February 1918 it was reported that Captain Falconer, under suspension since the tapping incident, had had his departmental trial before First Deputy Commissioner Leach and had been fined five days' pay and then reinstated. He was to resume command of his old precinct.[48]

Magistrate Joseph E. Corrigan, an independent candidate for the office of New York district attorney, charged in November 1917 that not only had his private wire been tapped but also that Tammany district leaders throughout the city were intimidating business men who had declared for his candidacy. "Last night I received word from the Police Department that my private wire had been tapped," he declared. As far as Corrigan was concerned it could only have been one of two men who were responsible—either Edward Swann or William Ransom.[49]

The committee on the amendment of the law of the New York City Bar Association disapproved of Assemblyman Leininger's bill making it a felony to tap telephone wires. It did so because it thought the bill was not rigorous enough. In a letter sent to Governor Whitman and members of the New York State Legislature the committee said that phone conversations should be as safe from eavesdropping as were letters, as a fundamental principle of civil liberties and constitutional rights. The Leininger bill would make it a felony to use a device to overhear phone conversations, except on the written approval of the U.S. district attorney or by order from an officer or member of the U.S. Secret Service. The committee urged "that no one should be permitted secretly or clandestinely to intercept conversations over the telephone."[50]

New York Governor Whitman announced on April 24, 1918, that on all bills that concerned the war he would follow closely the wishes of the authorities in Washington. Those Washington authorities caused Whitman to veto that day, he explained, the Murphy "listening in" bill that would have made it a misdemeanor to tap a telephone wire except upon the order of a Supreme Court justice. Whitman's action followed instruction from U.S. Secretary of State Lansing and U.S. Attorney General Gregory, who had sent word that any interference with listening in on the telephone would greatly handicap the work of the U.S. Secret Service.[51] Late in October of 1918 the United States Senate passed a bill wherein the tapping of telephone and telegraph wires under control of the Federal Government was made punishable by a fine of $5,000 or imprisonment for up to one year.[52]

In San Francisco on November 29, 1918, it was announced that District Attorney Charles M. Fickert would request the indictment of John B. Densmore, federal director general of the U.S. Department of Labor, who had made a report disclosing alleged irregularities in California prosecutions, including the case of Thomas J. Mooney. The indictment would be sought under statutes pertaining to wiretapping. The Densmore report, in connection with which, Fickert said, wiretapping was done, set forth many conversations alleged to have been carried on over telephone wires leading to the district attorney's office, Fickert's law office and other offices. A grand jury was to investigate the Densmore charges of irregularities in the prosecution of Mooney and others in connection with the bomb murders on Preparedness Day, July 22, 1916, in San Francisco.[53]

As a member of the U.S. Department of Labor, Densmore was sent to San Francisco to investigate the Mooney case and while there he tapped the telephone wires leading to the office of the district attorney who prosecuted Mooney. He also placed a dictograph in the office of the district attorney. The conversations he heard revealed a plot of corruption and bribery. Reportedly, the California Legislature was to conduct an investigation of the whole matter.[54]

Tom Mooney (1882–1942) was a socialist and a fighter for labor. He had briefly been a member of the Industrial Workers of the World (Wobblies) and after he moved to San Francisco he became the publisher in that city of *The Revolt,* a socialist newspaper. He was well known there as a militant and socialist. He was tried and convicted for the Preparedness Day bombing that resulted in 10 deaths and 40 injuries in the midst of the parade that day. The bombing took place at the height of anarchist violence in America. Mooney had earlier warned of rumors that agents provocateurs might try to disrupt the parade. Witnesses at the trial were coached by Fickert. Mooney was convicted and sentenced to hang. Later a commission established by U.S. President Woodrow Wilson found no evidence of Mooney's guilt, but despite this fact he remained in jail. In 1918 his sentence was changed to life imprisonment. Mooney was pardoned and released in 1939. Although Densmore's report was dated November 1, 1918, it did not become public until July 1919 when the conclusion that Mooney had not received full justice in his San Francisco trial was announced. Much of the information obtained by Densmore and his assistants, the report said, "was secured by the use of dictographs placed in the office of Charles M. Fickert, district attorney of San Francisco." More than 100 pages of conversations were recorded from those taps and it was reported that all of those conversations were overheard by two or more persons each time. When President Wilson wanted more information on the Mooney case he asked the secretary of labor to get it for him. Densmore, who had been solicitor for the Department of Labor and had just become director general of the U.S. Employment Service, went to San Francisco, looked over the ground, and decided on a deep and secret investigation, which included the use of dictographs.[55]

George and John Hesterberg, brothers of Henry Hesterberg, a Democratic leader and Brooklyn Superintendent of Highways, were arraigned in court in Flatbush, New York on March 15, 1919, on a charge of bookmaking. Those charges were the outcome of a raid on the café maintained by the brothers by police detectives who said they had tapped the telephone wires of that café and found them to be used for the placing of racing bets.[56] In court on March 28 the two brothers were discharged by Magistrate Folwell who held that sufficient evidence had not been obtained. Folwell refused to accept evidence obtained by police detectives who tapped the wires leading into the café operated by the brothers, declaring he could not hold the defendants merely upon the recognition of their voices over the telephone by detectives.[57]

A grand jury was investigating the city administration of New York City Mayor John Hylan in November 1919, giving particular attention to the Police Department. On November 9 it announced it had taken measures to establish the source responsible for the shadowing of its members. One of the jury's activities had been to delve into possible contacts between the New York Police Department and the underworld. One of the jurors, Raymond F. Almirall, said that jurors had ceased discussing jury matters with one another over their home phones as they were convinced those instruments were tapped. When they used phones to communicate with each other they used outside coin phones. Also it was said the grand jury room "had been looked over thoroughly for possible dictaphone systems." It was said that none were found. Several weeks later it was learned that certain messages sent to the grand jury had never arrived. Members of that jury were still worried about taps and there were still rumors circulating that listening devices had been installed in the jury room.[58]

Third Deputy Police Commissioner Augustus Drum Porter, indicted on charges of neglect of duty by the New York City grand jury in its inquiry into the alleged vice graft conspiracy in the New York Police Department, was removed from office on March 20, 1920, by Police Commissioner Richard Enright. Porter's indictment came as a result of the testimony of three detectives, formerly members of Inspector McDonald's special vice squad, to the effect that in raiding an apartment on the night of November 12, 1919, after they had tapped the telephone wire of the place, they found Porter in a room with a woman. Porter said it was all a frame-up. It all began back in September 1918 when two women were taken in a raid on a hotel. They claimed they were teachers at a dancing academy and that they were particular friends of Porter and that the officers who arrested them would be dismissed from the force. Policeman Fred Sorger, along with fellow officers Wheelwright and Cushing, eavesdropped on Adele Goodell's telephone from 6:00 to 9:00 p.m. the night of November 12, 1919, before the arrival of the man and woman said to be Porter and his woman companion. Police Lieutenant Sweeney gave Sorger the order to tap the telephone wire to the Goodell flat.[59]

Moonshiners operating in the Danville, Virginia, area were reported on June 18, 1920, to "have tapped telephone wires and learned of intended raids, hours and sometimes days, before State prohibition officers appeared on the scene," according to W. T. Shelton, in charge of drug enforcement activities in that area. Shelton thought that explained the inability of his agents to trap illicit liquor distillers. He said he once got a tip and went to what he had been led to believe was a still but that would be deserted when he arrived. When Shelton arrived he was met by the man he had been led to believe was the operator of the still. That man wore a gun and told Shelton to sit down so they could talk things over. Shelton said that man told him he had listened in on the rural wires when Shelton's informer had delivered the tip.[60]

Charged with collecting money under false pretenses, four men were arrested on November 5, 1920, in an office in a building on Broadway in New York City by two New York Police Department detectives, Maskiell and Maney, who had tapped the phone lines of the office and listened in as funds were solicited for allegedly charitable purposes. When the office was searched police found a list of 4,000 prominent men and business concerns with suggestions as to the amounts to be asked from each and the best way of approaching them. The New York Police Department has received numerous complaints from people swindled by these men and on October 26 they tapped the phone lines of the office. The scammers pretended to represent people and clubs that they had no association with.[61]

A 1920 sketch of New York City Third Police Commissioner Augustus Drum Porter. Due to evidence gathered from a wire tap Porter was removed from his post. He had been caught consorting with prostitutes.

Samuel Untermyer, counsel to the Lockwood Committee, sent to Acting Police Commissioner John Leach a protest in November 1920 against what he believed to be espionage against him. He said that the telephone wires running to his office and to his city and country houses had been tapped and that he was being followed by detectives from New York Police Department headquarters wherever he went. Leach denied that Untermyer's wires had been tapped, adding that the department had not tapped the wires of anyone connected with the investigation. The Lockwood Committee was established to investigate allegations of a conspiracy among the building trades of New York City. Many illegal acts were uncovered and many convictions resulted among the building trades unions.[62]

Thomas B. Donaldson, Insurance Commissioner of Pennsylvania, was held by a magistrate on November 7, 1921, in Philadelphia on $10,000 bail on charges of conspiracy that grew out of an investigation of private fire insurance adjusters under the direction of the Pennsylvania State Insurance Department. Donaldson "solicited" contributions from insurance companies to investigate an alleged arson ring. Harry W. Haggerty, special agent of a telephone company, testified that he had warned Donaldson that tapping of telephone lines was a violation of the law. The prosecution against Donaldson was launched by the complaint of J. Milton Young, an insurance adjustor whose phone wires were among those alleged to have been tapped.[63]

10

Dictograph: the First Bug

The world of electronic surveillance took a huge step forward with the invention of the dictograph (often misspelled as "dictagraph"; the present work uses "dictograph" throughout except in quoted matter). Tapping of telegraph wires required electrical skills and telegraphic skills. Especially difficult to master were the telegraphic skills. Wiretapping was not a profession that a person could easily master through self-education. Tapping of the telephone wires made the job of surveillance one step easier. Electrical skills were still necessary, but nothing beyond that. Dictographs made the process easier still. Electrical skills were not really needed, just the ability to break into a place, plant the bug, run the wires (all well-hidden, of course) out to the listening post and sit down and wait. During this period news stories often talked about conversations being recorded from these tapping endeavors. What was true of the period covered by this book was that no conversation was ever mechanically recorded, automatically. That is, all recording of overheard conversations involved a person sitting live in real time at the listening post and taking the words down verbatim—obviously that was done only by a trained stenographer. That conversations were not captured automatically was a problem for the eavesdroppers. Some went so far as to use two trained stenographers taking down the same conversation, for verification and honesty.

When the dictograph arrived on the scene in 1911, as a bug, it was also a time when the telegraph was fading into the background. Race results were transmitted over the telephone rather than the telegraph and there was little reason any longer to tap the telegraph lines. When the telephone arrived and quickly became a part of every business and commercial operation, and virtually every home not long after that, it expanded the field of targets for surveillance. The telegraph lines could be tapped only by highly skilled individuals; in any case, only the larger businesses and institutions had a private telegraph wire to tap. Individuals were mostly excluded. When the telephone arrived it dropped the skill level necessary to engage in surveillance and made almost everybody a potential target. Note that in the period covered by this book all telephone taps were effected outside of the targeted home or office. That is, the tap was made on the wires outside of the home and then the tap wires run to the listening post. No break-in of the home was effected or necessary and no interference took place with the physical telephone receiver. Dictographs made everything easier still in that the skill level dropped further (excluding the skill needed to break into a target house or business to plant the bug in such a way as for it to elude detection). The targets for surveillance increased again, to include those few without a telephone. The dictograph and the telephone tap complemented each other. While the dictograph allowed the eavesdropper to hear everything going on in a room, it included only one half of a phone conversation. The telephone tap took care of that omission. What perhaps limited the spread of the dictograph in the first decade of its existence was the added expense of having to have at least one stenographer on duty live at the listening post at all times. Nevertheless that first decade of the dictograph saw it receive a huge amount of media attention. All businessmen were urged to get one for their own office so they could record conversations they had with clients and other visitors. Such an accurate record was needed because it was, after all, a very duplicitous world.

It all began with Miller Reese Hutchinson (1876–1944), an American electrical engineer. He invented the first electrical hearing aid, called the Akoulathon (perhaps Akouphone), around 1895. In 1898 he patented an improved version called the Akoulophon. By 1902 he had refined the hearing aid into a more portable form powered by batteries, which he then was calling the Acousticon. In 1905 Hutchinson turned over the rights for the Acousticon to Kelly Monroe Turner (1859–1927). He patented an improved Acousticon in 1905, with that patent granted in 1907. Turner went on to improve the hearing aids and apply the technology to other products. One was the dictograph, which was an early hands-free inter-office intercom system. Turner's General Acoustic Company was renamed Dictograph Products Company because of the market success of the dictograph. As well Turner developed and marketed one of the first, if not the first, electric eavesdropping devices, which was also called the Dictograph, and was announced in 1910. Having the same name for two devices was confusing and sometimes the intercom machine was called the "commercial dictograph" while the bug device was called the "detective dictograph." However, such distinctions were not always used.[1]

Turner's device was first mentioned in the media in 1906 when he was developing both the intercom and the bug, probably simultaneously. In the end the technology was pretty much the same. For the commercial dictograph the item was in a standard black box that sat on a desk and was clearly visible. It was a bulky box as it had to have switches for each of the offices in the building the intercom was connected to. To use it as a bug was to strip the device done to its bare parts and hide it in a room. An account published in December 1906 noted that a little instrument had been demonstrated to a number of interested observers in Washington, D.C., at the U.S. Capitol on December 6. If it was put to general use members of the U.S. Congress could sit in their committee rooms, or in their offices in the new building that was then being erected, and hear what was being said on the floors of the House and the Senate. People who did not have

Miller Reese Hutchinson, 1902. His early invention of an electrical hearing aid laid the groundwork for the dictograph.

time to go to church, noted the reporter, might be able to sit at home and hear their favorite minister deliver his Sunday sermon, and patients in hospitals could have their spirits lifted by hearing music played and sung in halls or churches in remote parts of the city in their hospital rooms. The appliance that made all those possible: the "dictograph." Turner had been in Washington for a couple of days conducting a series of tests. Not only did the dictograph transmit sound, claimed the newsman, but it magnified sound. It was somewhat similar to the ordinary telephone in its general principle, but it was not necessary for one to talk directly into its mouthpiece in order to be heard at the other end of the wire. In testing his device in the House of Representatives Turner placed his transmitting apparatus, which was a disk "not larger than a small saucer" on Speaker Cannon's desk and led the wires out into the lobby. Someone then talked from a seat "some distance from the Speaker's desk. Despite the distance between the transmitting part of the dictograph and the Congressman who was speaking, persons stationed at the other end of the line out in the lobby heard him perfectly." Another benefit of the device remarked on was the fact that stenographers and clerks stationed in rooms served by the dictograph could thus take down speeches verbatim.[2]

Elliott Woods, the architect of the Capitol, planned in December 1906 to install an "acousticon" in the House and connect it with the new Congressional office building that was more than a block distant. By that instrument members would be able to sit in their offices in the new building and hear all the debates in the House. The acousticon was described as "a small black disc resembling the ear piece of a telephone, and about as big around as the end of a tomato can. The instrument is much more sensitive than the telephone and records and multiplies every sound." At the other end a listener could attend through a horn to a noise audible over the listening room or, by flipping a switch, listen through an ear piece.[3]

A commercial dictograph was installed in the office of the recorder of deeds in Washington on April 20, 1907. It consisted of a central station on the desk of Deputy Recorder Dutton and six substations, one in the private office of Recorder Darcy, one on the desk of the receiving clerk or cashier, one in the comparer room, one in the front record room and the other two in rooms used by copyists. Each substation was connected to the central station directly and by simply dropping a small lever it was possible to have communication between one of those substations and any other. Installations of the device were said to be pending in the office of the secretary of commerce, the secretary of labor, and the office of the postmaster general.[4]

Turner was touting his dictograph to supersede the ordinary private telephone installation "with which every great business house of the day is equipped." With a dictograph it was possible for department heads in a business to communicate directly with any department in the building for conversation or the dictation of letters, without requiring the members of the staff to leave their own rooms. The speaker did not have to speak directly into the dictograph but be any distance from the machine, say 15 feet, and still be heard perfectly. The device was described as a small black box, about 12 inches long by six inches wide, and four inches deep, with a row of switch-buttons along its base corresponding to the number of departments within the building. Said a reporter, "The advantages of the Turner system are obvious while, moreover, absolute privacy between the speaker and listener is insured, as the line cannot be tapped at any intermediate point."[5]

Needless to say, the dictograph as an intercom system went on to become a huge and lasting success. An article published on February 26, 1910, mentioned the dictograph as a bug for the first time. That device was among the exhibits at the electrical show in Philadelphia that ran from February 14 to February 26. According to a journalist it was "an invention that has been adopted by the United States secret service and by large banks and other institutions where espionage upon visitors is regarded as an advantage to those in charge." As such a secret listening device it was then "used by one of the best known financiers of the world in his office in Wall Street, New York. The diaphragm of the instrument in his office is concealed in an ink well which is secretly connected with a wire that carries the conversation between the financier and his callers to a distant room where a stenographer at a receiving station transcribes it in her notebook unknown to the visitor." The newsman added, "In a similar way the invention is being employed by Chief [John] Wilkie of the secret service and by the president of many of the largest banks and trust companies in the country and abroad."[6]

Four months later a journalist named Francis Phillips published a similar story about the dictograph. He began by mentioning the recent financial cataclysms in New York's financial district due to "unprincipled men and their representations to prominent men" in Wall Street. When such scams unfolded the public often were treated "to the uninspiring spectacle of having its distinguished citizens regaling each other with such commonplace epithets as 'He's a liar. I never told him any such thing.'" In other words, it was indeed a very duplicitous world and what could be more valuable than an accurate verbatim record of a conversation that took place in private between two of those eminent Wall Street financiers? One scandal mentioned by name was the Hocking Valley Crash. In that case, one of the parties to it had, unknown to his pool associates, a little device on his desk "known in every day commerce as a dictograph." It was connected by an unseen wire to an adjoining office wherein sat a stenographer. A silent push of a button by the man at his desk starting a private conversation, and the stenographer was alert and transcribing, receiver to her ear. According to Phillips' all callers to that man's office were treated alike. And there was never any suspicion on the part of the visitors. "Concealed in the silver filigree of a clock that innocently stood upon the center of the financier's desk was a dictograph transmitter. Fully five feet away from the financier, yet every whisper—conversation was generally in ordinary speaking tones—was clearly audible in the adjoining room." In the old days, noted the reporter, it was always one on one; it was always A's word against B's, and who was to be believed.[7]

Phillips went on to report that the same dictograph bug system was then in place in the U.S. Treasury Department, the U.S. Department of Justice, and at the U.S. Secret Service. "Visitors pass in and out of these departments every day whose every word is subject to instant recall should occasion ever make its production necessary." At the Secret Service, no one outside of Chief Wilkie and one or two trusted assistants knew where the device was located. The building in New York City at 26 Broadway,

A composite showing the evils of Wall Street—the spider and the fly. Thus, a need for financiers to protect themselves by secretly recording the conversations of their visitors. The two men in the top left are talking while a bug is hidden in the clock on the desk between them. At the same time the stenographer is transcribing the conversation, in another room.

which housed "the most powerful corporation in the world—Standard Oil—has more ears than it has windows or doors—highly sensitized ears that do not offend the vision by making their presence felt as you stroll through the maze of offices." Most of Standard Oil's deals were cleared through the National City Bank in Wall Street—known in the street as the Standard Oil bank: "Every department of the bank, which occupies the old United States custom house, covering an entire square block, is equipped with one of these eavesdropping machines ready for every emergency." Taking their cue from Standard Oil, declared Phillips, other companies such as the Westinghouse factories in Pittsburgh and the Cramp shipbuilding yards in Philadelphia had installed the device "throughout their establishments as an every day adjunct of their business." Many ordinary businessmen were said to have the machine in their offices to dictate letters to their clerical people so those women did not have to enter their offices. Governor Charles Deneen of Illinois had one in his executive office at Springfield, as had Governor Adolph Eberhart of Minnesota at the state capitol in St. Paul. Thus, corporation agents and lobbyists needed to "be mindful of what they said." In the new $2 million headquarters of the New York Police Department, implied Phillips, the cell area was full of dictographs. Phillips concluded his piece by declaring that in the financial district of New York it was then impossible to tell if the office you were visiting was or was not equipped with a dictograph.[8]

The first major case to capture media attention, involving the dictograph as a bug was an incident involving the Ohio Legislature. A newspaper story published on May 3, 1911, declared that evidence would be submitted to the grand jury later that week that would show certain members of the legislature were organized into gangs or groups of anywhere from three to eight members "who cooperated with one another in soliciting and securing bribes from corporations and men who were trying to influence legislation." Those graft solicitors were said to be bipartisan in that in getting graft in exchange for votes no distinction was made between Democrats and Republicans. It was reported that prosecutor Turner had stenographic records of a conversation that occurred in private detective Harrison's room in the Chittenden Hotel relative to the exchange of "business between house and senate crooks." In that conversation a senator was telling Harrison how they worked with certain house members and the "Burns dictagraph" transmitted all he said to Prosecutor Turner's stenographer. Another senator whom Harrison said he paid $200 for working in favor of a certain bill said as follows into the dictograph: "There's a couple of other members who always work and vote with me on all bills, and we split three ways all money we receive. I'd like to get some money from you for them." William Burns, head of the Burns Detective Agency, had been retained to ferret out the corruption in the Ohio Legislature and declared it was the most corrupt assembly he had ever been called upon to investigate. In the Ohio House all six members of the calendar committee—under much suspicion—resigned except for Dr. George B. Nye, who had private investigators arrested for offering him bribes. Said Burns, "Of all the bribery investigations with which I have been connected the evidence in this is the most conclusive, the most unquestionable, and was gotten by the most up-to-date methods." He added, "The Ohio legislators are the most persistent grafters I ever saw in my life. They were after it all the time and would take anything. From all I could learn they've been grafting every since they came to Columbus in January."[9]

Charges of wholesale bribery in the Ohio Legislature caused a huge political sensation in that state. On Sunday, April 30, 1911, three men were arrested on affidavits filed by Representative George B. Nye of Pike County, charging them with paying him money to influence legislation. Immediately following their arrest it was discovered that they were all private detectives of the Burns Detective Agency, who were looking for bribery among the legislators. As of early in May it was reported the general opinion prevailing was that Nye accepted the money as a bribe but later became

148 Wiretapping and Electronic Surveillance

The Ohio Legislature scandal, 1911. Rodney J. Diegel is on the left and a stenographer in a listening room is transcribing. Note the woman is listening through a fairly large horn amplifier—not an ear piece. The dictograph in the clock section was borrowed from another story.

suspicious of the men who paid him and in an effort to clear himself had the three men arrested. Adding fuel to that speculation was the fact that while he turned over the correct amount of money with which he had been bribed—$200—it was not the same bills. All those bribed by detectives received marked bills. Nye turned in $200 of unmarked bills. As a report noted, "The evidence in these cases was secured by a dictagraph. This is an instrument which was located in the [hotel] room of the detectives and which carried the sound of their talk to an adjoining room in which an official court stenographer was seated, who took down the entire conversation."[10]

Later in May a reporter remarked that in addition to the device being used in the Ohio case, the dictograph had been employed in one of Pitts-

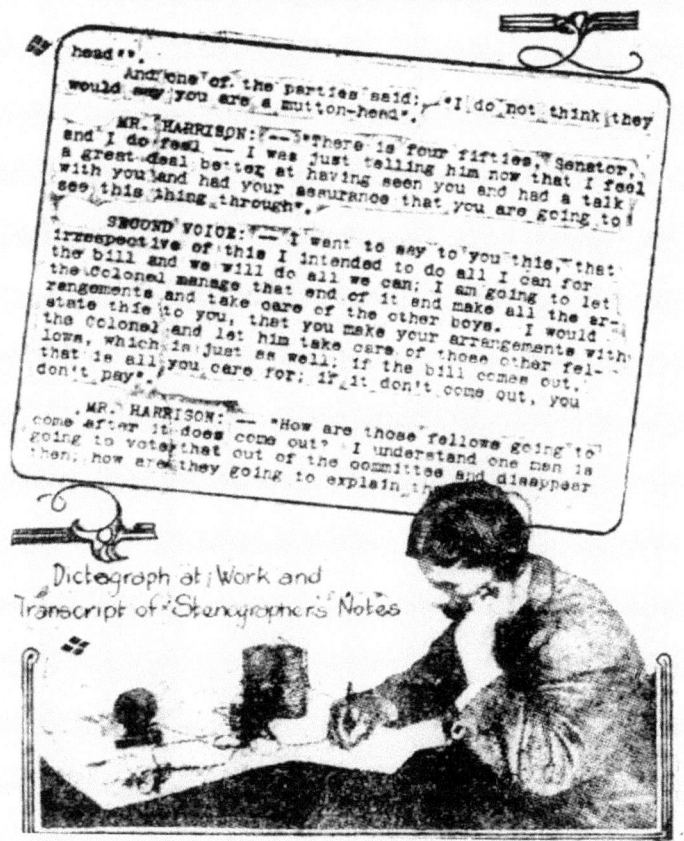

A stenographer at work in the 1911 Ohio Legislature scandal and a section of the transcribed and then typed notes, taken verbatim from the dictograph.

burgh's "most odorous" divorce cases.... It has been going on its silent, unostentatious way for some four years, ever since K. M. Turner, of Jamaica, L. I., invented it." The gist of this article was that no one was safe from eavesdropping anywhere. "And if there happens to be, under the table, or behind a picture on the wall, or back of a portiere [curtain across a doorway], a little disk of rubber and metal, that would slip into a man's pocket—why, then, a person sitting in another room can hold to his ear another rubber disk, no larger than a watch, and hear every word that is uttered, as plainly as if his ear were within two inches of the mouths of the people talking." Kelly Turner told the newsman; "I have always figured that the less said about the dictograph, in connection with detective work, the better. Now that Mr. Burns has told about using it, however, I have no objection to a little more publicity." Turner took from his pocket a complete dictograph outfit, which the reporter described: "Two disks, one about four inches across and an inch thick, the other the size of any ordinary watch. A battery which would shove into a vest pocket. That's all, except the necessary wire between transmitter and receiver." The larger disk was "planted" anywhere in the room. Two wires led away through the floor, walls, ceiling, or window.[11]

The dictograph in operation and inventor Kelly Turner, on the left.

Toward the end of June, 1911, Rodney J. Diegel, sergeant-at-arms of the Ohio Senate, had been indicted and was on trial for bribery. Diegel, it was alleged, called on detective Harrison at his hotel room and arranged a meeting for him with a number of Ohio senators to do a deal with regard to an insurance bill, and had then accepted money from Harrison for making those arrangements. All of that conversation, declared prosecutor Turner, had been captured by the stenographer from the dictograph listening post, in the hotel room located beside Harrison's room. Turner added a little humor to the proceedings when he said that when Diegel became suspicious and searched Harrison's hotel room before negotiations began, Diegel did not know what the dictograph instrument was when he saw it. Diegel did such things as look in the closet and under the bed, presumably looking for another person in the room who could bear witness. During Diegel's trial the dictograph was exhibited in court.[12]

What they had in this trial for bribery was that usual situation found in such cases, one person's word against that of another, the bribe taker and man offering the bribe. The defense objected strenuously to allowing the dictograph stenographic records to be admitted in court but, said a reporter, "the court ruled that it was corroborative of the direct evidence and was admissible." By September 1911 two convictions had been obtained in the case. Representative Nye had been tried and acquitted. In this case the dictograph had been hidden under a sofa in Harrison's hotel room.[13]

A decision from the Ohio Supreme Court in which the use of the dictograph in convicting corrupt legislators was upheld prompted an editorial on the subject. The editor noted that the state's case would have failed but for the admission of the dictograph evidence, which confirmed the claim that there was a bribery conspiracy and that Diegel was the go-between. He went on to conclude, "The dictagraph is here to stay—both in the hands of detectives and by formal judicial decree. The history of its achievements in a brief six months shows that invention had added new protection to the community and a new terror for crooks."[14]

Kelly Turner was said to have been summoned to Columbus, Ohio to testify in the trial of State Senator L. R. Andrews, one of the Ohio legislators indicted on charges of bribery. The prosecution had the same evidence at its command as it used for Diegel, who had been convicted and started serving his three-year term in the penitentiary around the beginning of March 1912. All the other trials in this case had been delayed while Diegel fought his conviction up to the state Supreme Court.[15]

One commentator thought the Supreme Court ruling that upheld the use of the dictograph evidence in the Diegel case had huge implications. Although he was a victim of a police "plant" Diegel actually convicted himself, argued the piece. He talked voluntarily and therefore could not take refuge behind the legal privilege of the man who is compelled to testify against himself. There was no compulsion in this case as Diegel talked freely. Diegel had first examined the hotel room, but he was looking for a hidden person and therefore checked under the bed, under the couch and in the closet. Diegel had been suspicious as he met a man [private detective Harrison] he thought was a lobbyist. This article pointed out that identification of the voice would likely play a key role in future trials that relied on dictograph evidence. There was also the possibility of faking transcriptions and notes and the importance of having more than one stenographer transcribe with the optimum condition being to have each stenographer alone in a separate room which creating the transcription. Those in favor of the device said that it was a safe and more humane expedient than the police "third degree." Put your prisoner in a cell, they argued, and let his friends visit him—the concealed bug would catch it all.[16]

In December 1912 Ohio State Senator George K. Cetone of Dayton was found guilty of accepting a $200 bribe from a Burns operative posing as a lobbyist in the spring of 1911. He was the third politician to be convicted in the assembly bribery case. Two representatives and Diegel had been convicted heretofore. In this case the jury deliberated for only 70 minutes. With the conviction of Cetone the last of the dictograph cases was disposed of. Other bribery indictments were still outstanding but none of them involved the dictograph.[17]

In Gary, Indiana, on September 8, 1911, Mayor Thomas E. Knotts, five of the nine members of the Gary City Council, City Engineer W. A. Williston, and a son of one of the aldermen, were arrested on charges of accepting and soliciting bribes in a heating franchise deal. The arrests were made on a complaint by T. B. Dean of Richmond, Kentucky, to whom that heating franchise was granted. Dean said he had given evidence of the attempted bribery to attorneys in Chicago before any money was transferred. A dictograph was said by Dean to have been placed in his hotel room and one was surreptitiously placed in Mayor Knotts' office. Records from those devices were expected to be used in the prosecution of city officials. Knotts had been arrested in the previous May on charges of embezzlement, perjury and malfeasance in office. Those charges were dismissed. Dean charged Knotts with having received $5,000 as his share of the bribery deal that saw Dean be awarded the heating contract. Amounts Dean said he had given to the other men involved in the scam ranged from $250 to $2,000.[18]

As part of the set-up in the Gary case, Dean was searched, reportedly, by four men before entering Knotts' office and then searched again by the same men as soon as he left that office—to show he had the bribe money on his person going into the office but not on the way out. Those four were police deputies. An excited Dean told those deputies just where in the mayor's office they would find that envelope. Dean told authorities how he came to the city to try and secure a heating contract. Before long, though, he said he understood he would have to pay in order to win that contract. After he consulted with his lawyers he arranged the set-up. It was reported that Knotts arrived in Gary "broke" some four years earlier, and lived in a "rude hut" on the sand wastes. "Today he is rated a millionaire and lives in one of Gary's finest homes."[19]

Gary, Indiana, Mayor Thomas Knotts, 1911. Accused of bribery from evidence obtained by a dictograph. However, it all proved to have been faked evidence. It was part of a darker side of the dictograph—the ability to use it to construct false evidence—that got little publicity.

Then the whole story came apart. In February 1912, it was reported that the story was a frame-up. Thomas Dean himself, it was said, dictated the so-called dictograph records which the prosecution of the Gary bribery case hoped to use to convict and drive out of office a number of Gary city officials and politicians. Meyer Himmelblau was the stenographer employed by Dean to do the transcribing but he came forward to confess that part he played. He told how Dean dictated the alleged records of conversations and from which the stenographer created the shorthand notes to match those forged conversations. Himmelblau added that the money paid him had been received from the Burns Detective Agency. Dean had been indicted for perjury.[20]

According to Himmelblau he came forward to confess as he was conscience-stricken after an official had been sentenced to the penitentiary. He swore an affidavit in Chicago that Dean had made him destroy a stenographic record of a conversation between Dean and John Nyhoff, a member of the Gary Board of Public Works, dictated a new record and had Himmelblau make shorthand notes of the "faked" record. Records of conversations with seven other people were made the same way. Before the stenographer confessed, two of the nine men named had been tried. One of them, Alderman Walter Gibson, was convicted and sentenced to from two to 14 years in penitentiary. The jury was hung in the case of the second man, C. A. Williston, City Engineer. Those two trials cost a reported $25,000. Since he had signed the affidavit confessing his part in the story and had perjured himself, no trace of Himmelblau could be found. Speculation was that he had been "spirited away by someone interested in the case." There were no more published reports on Dean or Himmelblau.[21]

On October 27, 1911, an article remarked that a scandal had hit Kentucky politics and threatened to overthrow the carefully laid plans of Democrats "to buy the forthcoming gubernatorial and legislative elections." It was learned at Lexington that D. S. Johnson, manager of the Johnson Detective Agency of Cincinnati, had a score of his detectives working throughout the state over the previous two weeks gathering evidence and affidavits to be presented in half a dozen cases. Some time earlier, it was said, it became known that Democratic leaders were desperate at the lack of enthusiasm for the McCreary ticket and that it was decided to buy 200 black votes in every town of over 5,000 in the state of Kentucky. Through the use of black spies and the dictograph which was, noted the journalist, "so startlingly and effectively used in the Legislature graft cases in Columbus O.," detective Johnson succeeding in securing confessions from a number of "negroes who had sold their voting certificates." And, continued the report, "When confronted with their own words taken down by a stenographer at the other end of the dictograph, many of the negroes broke down and confessed, implicating men high in the councils of Democracy in the town affected." The prevailing price paid to a black man for his voting certificate was $3. One man arrested was H. S. Sarasohn. He bought the voting certificate from a black man named George Ross for $3 and had that certificate in his pocket when he was arrested. James B. McCreary (1838–1918) was governor of Kentucky from 1875 to 1879. Then he served in the U.S. House of Representatives from 1885 to 1897 before serving as U.S. senator from Kentucky from 1903 to 1909. McCreary served a second term as Kentucky governor from 1911 to 1915. There were no more reports about Sarasohn.[22]

The Los Angeles Times building in Los Angeles was bombed on October 1, 1910, by a union member belonging to the International Association of Bridge and Structural Iron Workers (headquartered in that city). That bomb explosion started a fire and 21 employees of the newspaper were killed with another 100 being injured. Brothers John J. and James B. McNamara were arrested in April 1911 for the bombing. James admitted setting the explosive, was convicted, and was sentenced to life in prison. John was sentenced to 15 years for bombing a local iron manufacturing plant in the area. He returned to the Iron Workers union as an organizer. Around that time employers used labor spies, agents provocateurs, private detective agencies, and strike breakers as they engaged in a rigorous campaign of union busting. Local, state and federal law enforcement agencies cooperated in that campaign. Responding to that campaign, the Iron Workers elected Frank M. Ryan as president and John McNamara as secretary-treasurer in 1905. Unions everywhere in America were being weakened in the time period around 1910, from the relentless campaign waged by employer and state against them. Beginning in late 1906, national and local officials of the Iron Workers launched a dynamiting campaign, with a goal of forcing the employers to the bargaining table—not to destroy plants or to kill people. Between 1906 and 1911 the Iron Workers, reportedly, blew up 110 iron works, but with minimal damage amounting to only a few thousands of dollars in total. The National Erectors Association, a coalition of steel and iron industry employers, was not unaware of who was responsible for the bombings because Herbert S. Hockin, a member of the Iron Workers executive board, was a paid spy for the Association. Weeks passed in Los Angeles after the bombing with no arrests. On December 25, 1910, a bomb went off at the Llewellyn Iron Works (for which John McNamara would be convicted), partly wrecking the plant. In the wake of the bombing the city of Los Angeles posted a $25,000 reward, an employer group posted another $50,000 and the city of Los Angeles hired the Burns Detective Agency to investigate. The Burns agency had been investigating some of the iron manufacturing plant bombings for the previous four years on behalf of the National Erectors Association. From Hockin, in his position as a paid spy within the union, Burns heard that Iron Worker member Ortie McManigal had been handling the Iron Worker bombing campaigns on orders from Ryan and secretary-treasurer John McNamara. On April 14, 1911, Burns and some of his other operatives arrested McManigal and James McNamara in Detroit, took them back to Chicago and held them incognito for a week, April 13–20, in the private home of a Chicago police sergeant by the name of William Reed. McManigal reportedly confessed, told what he knew and implicated Ryan, John McNamara, Hockin and others within the Iron Workers leadership. Burns arrested John in Indianapolis and went before a judge who released John into Burns' custody even though he had no legal right to do so. Within about 30 minutes after his appearance before the judge John was on his way to California, in the control of the Burns Detective Agency. James and McManigal were also California-bound. All of them arrived in Los Angeles on April 26, 1911. William Burns had engaged in kidnapping, misrepresentation of his status as a law enforcement officer, unlawful imprisonment and possibly torture in the handling of James and McManigal. Iron Worker President Ryan asked Clarence Darrow to defend the McNamara brothers. Initially he refused but then agreed after American Federation of Labor chief Samuel Gompers stepped in to make a special plea. On May 5, 1911, the brothers were arraigned and pled not guilty. McManigal turned state's evidence and was not charged. The defense was weakened further on November 28 when Darrow was accused of attempted bribery of a juror in that trial. Darrow was acquitted in his first trial on that charge of bribing a juror. When charges were brought against him in the second bribery case the trial ended in a hung jury. James B. McNamara died of cancer in San Quentin on March 9, 1941. John J. McNamara died in Butte, Montana, on May 8, 1941; he was then working as an organizer for the Iron Workers. Bugging was involved at every step of the way in these cases. The jail cells occupied by the McNamara brothers were bugged, Frank Ryan's union office in Indianapolis was bugged, and a hotel room used by a lawyer working with Darrow, and wherein Darrow met that lawyer, was bugged.[23]

Miss Eula Hitchcock of Los Angeles was a stenographer employed by Sam Brown, chief detective for the state of California in the McNamara case. In December 1911 she was reported to be largely responsible for the Darrow bribery investigation that resulted in the arrest of Bert H. Franklin (the man who implicated Darrow). Eula was a confidential assistant to Brown who heard of an alleged plan to "fix" the jury and then told her superior about that plan. A dictograph was placed in the cell occupied by the McNamaras; Eula, stationed at the other end of the wire, overheard conversations that led to the arrest of Franklin. Her story prompted a comment from a California newspaper: "Among the wonder instruments perfected the past decade none stands out more striking at the same time so dreadful and uncanny, as the Dictograph," remarked the editor. "During the past two years this little instrument has been responsible for the culmination of many noted criminal cases, its most important one, by long odds, being the Los Angeles dynamiting cases, which so suddenly came to a partial termination by the confession of the McNamara Bros." The National Dictograph Company of Jamaica, Long Island, New York, had just opened an office in San Francisco at 685 Market Street in the Monadnock Building (Room 558). The piece continued by declaring, "No investigator or detective should be without one of these instruments, as its worth can be easily demonstrated as of great value in the many private cases given detective agencies to handle. Descriptive folders, prices and fullest data can be obtained by writing or calling." While this piece looked like an ordinary news article it was, obviously, a plant itself, a company-written press release that tried to mask itself in the newspaper as news and/or comment.[24]

An article published on May 21, 1912, included a photo of William Burns and enthused, "The greatest detective is shown holding the instrument which is now the terror of all lawbreakers, the dictagraph." The piece added, "It is generally believed that the McNamara brothers decided to plead guilty to the blowing up of the Los Angeles Times and other dynamite plots after they had discovered dictagraph receivers in their cells." And, it was reported; "The instruments had been placed in the cells by Detective Burns and were connected by wires with rooms in another part of the jail where Burns, and members of the prosecution secured an appalling amount of evidence by listening to confidential talks between the McNamaras and their lawyers."[25]

A report published on February 17, 1912, brought the news that the federal government had reportedly secured the conversation of President of the bridgeworkers union from a dictograph that had been placed in the office of the ironworkers.[26]

It transpired that whatever Frank M. Ryan and other indicted union officials said in their office about the dynamite conspiracy from October 1911 until February 1912 in their Indianapolis, Indiana, offices was alleged to have been learned by the government through a dictograph that had been planted at the headquarters of the International Association of Bridge and Structural Ironworkers. The bug had been discovered on February 17 and had been hidden under the drawer of a desk used by union president Ryan, Herbert Hockin (secretary and treasurer), J. T. Butler (vice president), and others. Around that table those men had conferred on their defense, pleas, and so forth, for upcoming legal cases. That enabled two government stenographers in the room below to take daily reports of such conversations. The use of

A 1912 shot of William Burns holding a dictograph.

the device was disclosed on February 17 when the government decided it was no longer of value because a woman clerk in Ryan's office was heard to say, "Well, I suppose they are hearing now whatever we say." That and other statements convinced Assistant District Attorney Clarence Nichols that the device was no longer useful. District Attorney Charles W. Miller said the device had worked satisfactorily for months and that many volumes of stenographic notes were taken and would be used at upcoming trials. In one of the last conversations heard, Ryan and Hockin discussed if either knew what a dictograph looked like and whether or not to try and go somewhere where they might see one—to know what to look for. A visitor to Ryan's office told him a dictograph was planted in the office. An examination of Ryan's desk then led to the discovery of the bug. Wires ran back from Ryan's desk through the floor into the room below, which was rented by the federal government. Here sat every day two stenographers under the direction of Rowland Evans, stenographer for the federal court, with a receiving apparatus to their ears. Those stenographers worked in relays, except at such time as it was desired to make a double record for corroboration. Both Ryan and Hockin said they were angered that such a device had been in the office. Said Hockin, "I suspected some eavesdropping scheme, but at first thought that the telephone wires might have been tapped." Yet later that year when Ryan and many others stood trial, it was revealed by District Attorney Miller that Hockin knew all the time the dictograph was hidden beneath Ryan's desk. Said Miller to the jury, "Hockin knew for two months that the dictagraph was there, because he helped put it there." Reportedly, the other defendants were inclined to discredit such allegations. Miller added that Hockin had delivered the keys to the Iron Workers offices to an outsider so that access could be had to the premises for the planting of the bug.[27]

Later that month, in connection with the Ryan case, it was reported that most of the lawyers connected with the case believed the records of conversations obtained from the dictograph would be admissible evidence in the trials of the indicted labor officials, who were charged with being connected in the alleged dynamite conspiracy. Each day that conversations from the Iron Workers tap were transcribed produced from 15 to 60 typewritten pages. Later a total of 38 labor union leaders were convicted with respect to the dynamite conspiracy. Frank Ryan drew the heaviest sentence, seven years, while the others received, for the most part, sentences of one to six years.[28]

Clarence Darrow, chief counsel for the McNamara brothers, was indicted in March 1912 on a charge of bribing the McNamara jury. He was reported to have been trapped by the dictograph, according to the state-

A 1912 shot of the desk of Iron Workers President Frank Ryan, showing where the bug was located, and a close-up of the dictograph.

ment made in Indianapolis on March 13 by Walter Drew, counsel for the National Erectors' Association. Drew declared the device had recorded conversations between Darrow and his associate counsel, John R. Harrington, which would be used in the trial in Indianapolis of the alleged dynamiters. Drew stated that a dictograph in the room in the Los Angeles hotel occupied by Harrington was used to record conversations relative to Darrow's defense.[29]

On March 14 Walter Drew admitted that Robert J. Foster, a private detective employed by the National Erectors' Association, the man who placed the dictograph in the office of Frank Ryan, had also recently placed a dictograph in a Los Angeles hotel room so that conversations between Darrow and one of his attorneys, Harrington, were overheard. With respect to those conversations, said Drew, "The chief value will be in the Darrow case, but some of the information obtained is of value in the dynamite case." Foster had left Indianapolis for Los Angeles some six weeks earlier and upon his arrival rented a hotel room that was next to the one occupied by Harrington. Also some six weeks earlier Harrington had been in Indianapolis to discuss the dynamite case with some of the ironworkers. Darrow and Harrington discussed some of those matters in Harrington's hotel room. Apparently the two attorneys suspected there might be some type of eavesdropping device because Harrington several times made a search of his room. However, he never discovered the bug and it remained

Left to right, Leo Longley, Robert J. Foster, and Waldo Faloon. Two stenographers and a detective transcribing with the dictograph showing how it was concealed behind a calendar in the hotel room of an attorney visited by Clarence Darrow. In that hotel room the two lawyers discussed the dynamiting cases they were involved with.

in his room for several weeks until a bellboy happened to find it. Then it was removed. Foster had two stenographers at work transcribing the conversations. Reportedly, the records concerning Darrow's case were turned over to the authorities in Los Angeles.[30]

Later newspaper stories showed photographs of Foster and the two stenographers taking notes. Those photos, it was said, would be introduced by the state as evidence to show that Foster really succeeded in "planting" the bug in the hotel, namely the Hayward Hotel.[31]

In Washington, D.C., on January 31, 1912, an attempt was made to impeach the testimony of C. G. McGowan, a Hines-Lorimer witness who testified that he did not hear W. C. Weihe make a statement concerning the $100,000 Lorimer fund, resulting in a stormy session in the Senate committee investigating the election of Senator William Lorimer of Illinois. William J. Burns of the Burns Detective Agency was called to the stand. It had been planned by him to lay the groundwork for his operatives to present "proof" of McGowan's having perjured himself. Items of "proof" listed were a letter and "admissions" McGowan was alleged to have made at Toronto, Ontario, to detectives, which were taken down by dictograph.[32]

One day later another session was held by the committee to show how the dictograph was used to procure an alleged admission from Charles McGowan, the Hines-Lorimer witness, that he perjured himself when he swore he did not hear C. F. Weihe tell of a Lorimer election fund. It was explained by private detective A. C. Bailey, a Burns operative. For nearly two hours Bailey read from "notes" made from day to day of remarks that McGowan was alleged to have made to Bailey, posing as a claims adjuster of the American Bridge Company. The dictograph was placed in a hotel room prepared for McGowan.[33]

The senate committee investigating the Lorimer controversy met again in late February 1912, at which time the stenographer's report, taken from a dictograph, was brought into question. Said a reporter, "The stenographer notes of the Burns' stenographer have been shown to have been largely faked." An experiment was conducted and the stenographer was given an opportunity to demonstrate his ability in the same manner that he claimed he took the notes which were introduced as evidence. He failed the test, showing that his former story was not in accordance with the facts. The reporter concluded by remarking, "A number of stenographers corroborated the statement of the official stenographer of the committee that the notes taken by the detective's stenographer were faked."[34]

Despite the less-than-impressive record of dictograph usage, as reported above, the device took the media and the public by storm. An editorial cartoon published in the *Tacoma Times* on February 10, 1912, showed, in humorous fashion, how hard it was to conduct a conversation that could not be overheard, and the lengths one had to go to in order to attain some privacy.[35]

A lengthy article in praise of the device appeared in a Chicago newspaper in February 1912 and was reprinted in various other papers. The piece was one of several that would appear in the 1912–1913 period and purported to show how pervasive the device was in society. It began with the wildly exaggerated statement that 800 dictographs were then at work

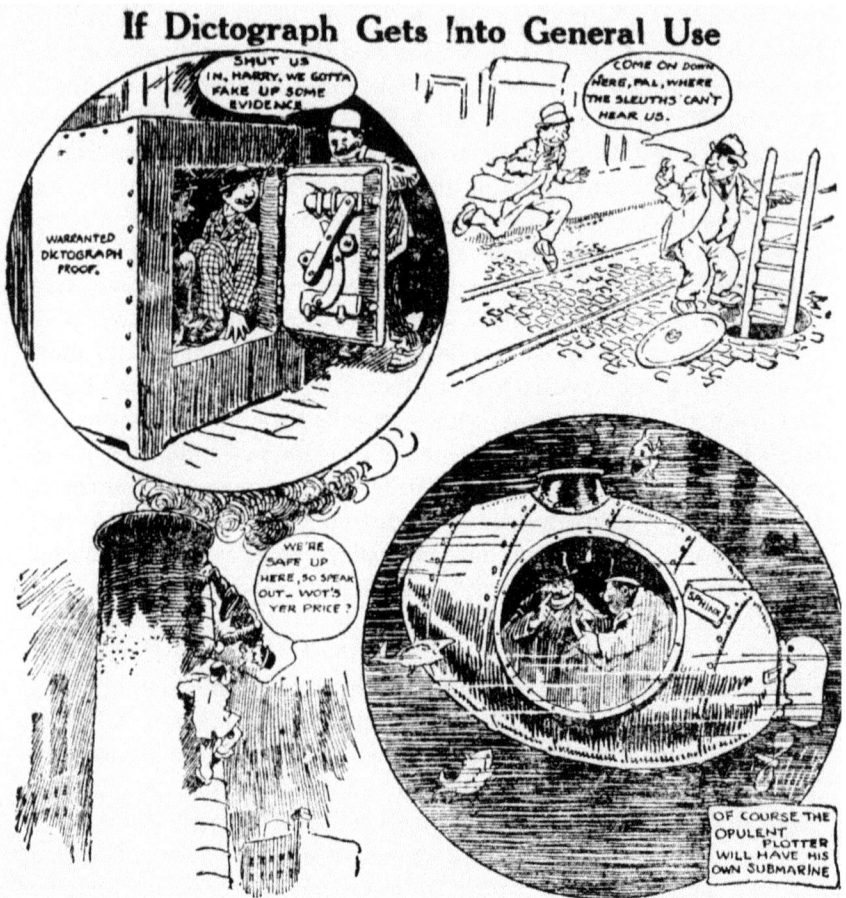

A humorous 1912 editorial cartoon illustrating how hard it would be to conduct private conversations and the lengths to which people would have to go to attain that end, if the dictograph became as pervasive as some predicted.

in Chicago. The reporter who wrote the piece said he went to the offices of the William J. Burns Detective Agency. At that agency they told the reporter that all the dictographs the agency had were then in use: "We use the dictograph in all our big cases now," declared Raymond J. Burns, superintendent of the Chicago office. "We find it indispensable. This instrument was first used by us two years ago in the Illinois Central graft cases. That was the first time it was ever brought into play this way. Before that it had been used as a desk contrivance in connection with telephones for the purpose of dictation." He added, "We used it again with success in the Columbus legislature bribery case. We used it in Gary and we used it in

Toronto [two cases that had been discredited as fakes]. But these cases are by no means all. Those are just the big instances. I could name many others if I had the time. It's certainly a wonderful little contrivance and makes our work a lot easier." The reporter then went to the office of a company that manufactured dictographs "along with such innocent things as aids to the deaf and office conveniences." (This would have been Turner's plant, since it was the only one that made the device.) The manager at that facility picked up a disk about four inches in diameter. He hid the disk by hanging it on a hook and covering it with a calendar. Then he ran the wires through a partition into the adjoining room and connected them with a battery and receiver. The whole instrument—transmitter, receiver, and battery—could be held easily in the palm of a person's hand, enthused the journalist. The transmitter consisted of a rubber "sound collector." Within that was a disk of aluminum the size of a watch in which there was a delicate carbon membrane. Behind the membrane were six tiny electrons containing minute carbon balls, which were set vibrating violently at the smallest sound. The vibration of the carbon disk was stimulated and that vibration was transmitted over the wires by a weakening and strengthening of the current, as in a telephone. At the other end was a receiver—again as in a telephone—and a battery. The operative in a detective case sat with the receiver clamped to his ear and by means of shorthand reproduced the conversation going on in the other room. According to this manager, "There are 800 of these machines being rented out in Chicago now. We have to be careful in renting them to make inquiries for they may easily be used in a harmful manner. We never let a machine go out without a careful investigation of the party desiring the service." He added, "Most of the important detective agencies have them in use and they are used for other purposes—some mighty peculiar ones, too. They would make good stories if I could tell them."[36]

Another article about the pervasiveness of the device appeared in a Washington, D.C., newspaper on February 18, 1912. It talked about the introduction of the dictograph into Washington where, "as a few people know, it has already entered the offices of more than one prominent financier...." All the executive had to do was press a button on his desk. That caused a buzzer to sound in the next room and the stenographer there on stand-by stuck the apparatus in his ear and recorded the conversation. The machine was so precise that it could capture even a whisper. Lawyers were also said to be big users of the device, using them, for example, when the husband in a divorce case visited the wife's attorney and tried to bribe him. It had reached a point, said the article, where some visitors to bankers wrote down on a piece of paper what they wanted to say and when it was read took it back and destroyed it. The journalist speculated that as the cost of the device became cheaper, ordinary homeowners would take to installing them. Thus, Mrs. Smith could send her visitors Mrs. Brown and Mrs. Jones into a room where the dictograph was planted and then listen to them discuss herself. "How extensive will ultimately become the use of the dictograph no man is not able to foretell," wrote the reporter. "But with its advent into Washington there seems small doubt but what it will become the most valued and dreaded of implements."[37]

Every word is carefully saved and may become Exhibit A.

Another 1912 humorous sketch about the growing use of the dictograph by businessmen and professionals who wished to record every word spoken by visitors to their offices, unbeknownst to those visitors.

It was reported from Pittsburgh on February 23, 1912, that when Andrew W. Mellon's divorce suit against his wife, Nora McMullen Mellon, was heard it would be the first time the dictograph had ever been used in a divorce suit. Mellon, worth $30 million, had had his home equipped with the device more than a year earlier. Since she became a defendant in the divorce suit, Mrs. Mellon had received $15,000 a year in alimony and the sympathy of those "in her set, who resented the means to which her husband resorted to get evidence against her." It was rumored that a servant informed Mrs. Mellon of the device. However, due to the bug, a co-respondent had been identified and named in the suit.[38]

Another long piece on the rapid spread of the device was published in a New York City newspaper on February 25, 1912, with an emphasis on the idea that the scientific eavesdropper brought a new terror to evildoers. It began by observing that malefactors would have to think twice about all conversations they had and check all furnishings, attachments, and so forth, in their meeting places. In the previous six months, said the newsman, the device had come into "sensational prominence." It had tapped the secrets of prison cells and revealed conspiracies in hotel rooms and offices. Also mentioned were union leader Ryan, the Columbus Ohio Legislature case, the Lorimer case and the Gary, Indiana, case. William J. Burns was called the first person to see the dictograph's possibilities in detection work and the first in the area to use it that way. The *New York Tribune* reporter who wrote the story went to see the inventor and manufacturer Kelly Turner at his Broadway office within New York City's theatrical district. (The factory itself was on Long Island.) Therein Turner was called the inventor of "the Acousticon, the interior telephone and the dictograph." The acousticon, said the account, had been used for some years in churches and theaters to enable the deaf to hear, noting, "All these devices are nearly related. They might be called merely superlative telephones." They magnified sound and transmitted it over a wire. The reporter made the distinction between the "commercial dictograph or Turner interior telephone [intercom]," while calling "the detective dictograph ... an adaptation of the interior telephone for detective work." The difference between them was basically two-way communication versus one-way. It was reported that Turner had a difficult time getting anyone to take his "ear machine" seriously for detective purposes. At the time of the Harry Thaw murder trial (he was tried twice, in 1907 and 1908), Turner's suggestion that his device be utilized was turned down by an assistant district attorney. Then he crossed the ocean and showed a form of his invention to the late King Edward of England, who thought that Scotland Yard should have it. The police of Paris and other cities took up the dictograph before anyone in America did, said the article: "The European crime hunters used it and kept mum. There are no noisy newspaper men over there to tell about such things." No other evidence appears to exist to confirm the device was used by the Europeans.[39]

According to this piece, the U.S. Secret Service was the first to get wise to the device in the United States when a man named "Billy" Burns began to dabble with the machine. William Burns had been with the U.S. Secret Service before he established his own detective agency. The reporter recounted one story about two Italian crooks who were placed in a Pennsylvania jail cell. A dictograph was planted in their cell. Standing by was a stenographer waiting for them to talk. They said nothing for five days but on the sixth day they started to talk and it was all recorded by that stenographer. Another use for the machine was in extracting confessions. The interrogator did not need a third party in the room. He could take the subject into a room alone, gain his confidence and urge him to tell what had happened. "He will talk because just the 2 of you are there. The dictograph will capture it all, through the stenographer." Turner explained to the reporter that in the right hands, the instrument had a "moral effect," but in a few cases it had been turned to "immoral and pernicious uses" such as for blackmail. For that reason, said Turner, the device was not sold or rented to "irresponsible parties." It had been vainly asked for by men who would not give their names and "were presumably crooks, or persons with an immoral design." The instrument was supplied to "accredited detective agencies and government officials. An honest business man or a respectable corporation can have it. It is sold for $150 or rented for $50 a year." (All other accounts said that purchase of the machine was not possible to anyone—only rental was allowed. No other source ever mentioned a price for the device.) Turner noted that a mechanic would install the device, for those who required it, "but any intelligent person can do that himself." The only "apparent flaw" in the dictograph, Turner grudgingly admitted, was the reliance upon a stenographer to transcribe the conversations. "In the Lorimer case, and more recently the Gary, Ind., case, doubt has been cast on the accuracy of the stenographic reports," observed the story. It was possible, Turner pointed out, to have two stenographers simultaneously recording, but that an ideal solution would be to make the instrument "independent of human fallibility." To that end, attempts had been made to attach the dictograph to a phonograph or dic-

Opposite: **Several 1912 photographs illustrating the rapid spread of the bug as it captured a large amount of media attention. Shown are a stenographer at work, a private detective planting a bug, and device inventor Kelly Turner depicted below a hand displaying a dictograph.**

 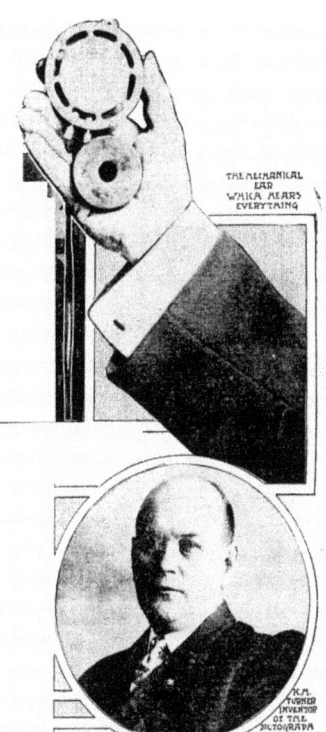

RECORDING CONVERSATION TRANSMITTED BY MEANS OF DICTO-
GRAPH.

THE MECHANICAL EAR WHICH HEARS EVERYTHING

K.M. TURNER INVENTOR OF THE DICTOGRAPH

HOW A DICTOGRAPH MAY BE HIDDEN TO TRANSMIT CONVERSATION

taphone so that a mechanical record of conversations might be automatically made. To that point, there had been no success, nor would there be in the time period covered by this book. The longest distance over which the dictograph had been used in a detective case was three miles, in a Western case, although no details were provided. A regular copper telephone wire was employed in that case. There was, said Turner, no practical limit on the length or distance between the device's transmitter and the listening post but an exclusive wire was necessary to avoid interruption. The device could be attached to any telephone wire. However, the telephone companies prohibited the use of devices they did not supply to their subscribers. Besides its use by the Burns Detective Agency, the Pinkerton Detective Agency, the U.S. Secret Service, the U.S. Department of Justice, the Army and Navy departments, and the New York Police Department headquarters, the dictograph was installed in the offices of the National City Bank, the Standard Oil Company, in railroad offices, and in many Wall Street establishments.[40]

In March of 1912 it was reported that the New York Police Department was getting information from and about the gang then under arrest for the Church Street taxicab robbery. Dictographs were planted in cells at police headquarters and detectives at the receiving end recorded conversations between the prisoners. By means of those devices it was learned that the original plan for holding up the East River National Bank messenger included shooting Geno Montani, the chauffeur who drove the bank taxicab, in the leg. His protests at that part of the plan caused the gang to change plans with one getting in the front seat of the cab with Montani and pretending to hold a gun on him. It was a $25,000 taxicab holdup and the dictograph evidence resulted in several indictments handed down by the grand jury.[41]

Another general article in praise of the machine appeared on April 7, 1912, and began by saying, "It is the first class eavesdropper of the age, its terror for criminals is not only in what it actually reveals to the sleuths, but in the fact the hunted criminal never knows just how much his pursuer has learned through this means, and he may know everything. Other traps have been invented from time to time, but the dictograph is about the cutest of them all." The entire apparatus weighed between six and eight ounces. Turner, it was said, had hit upon the idea for it several years earlier and by 1907 he had it in shape for laboratory work. Also noted was that it was an offshoot of the regular commercial dictograph [intercom] that Turner also manufactured. It was not for sale, but could be obtained on lease only, because of "its potential evil possibilities if it falls into the wrong hands." William Burns, it was said herein, got the idea of the utility of the instrument as a criminal catcher when he was a member of the U.S. Secret Service.[42]

One of the earliest cases to use the device involved the Illinois Central Railroad, according to an April 1912 article about an incident some years earlier. The management of that railroad became convinced that it was the victim of an extensive swindle in the matter of car repairs, although the railroad's own investigative people could not find any evidence. Then an outside private detective agency was called in, but it also found no evidence. When the case was referred to a second private detective agency (Burns) its operative concealed a dictograph in his room at a hotel with the receiver in an adjoining room. Summoning two of the suspected men to his room, he said enough to create the impression that he knew a lot about the suspected car graft. Then he excused himself and walked out. He went to the room next door where he picked up the dictograph receiver and listened to the two men incriminate themselves. They talked enough to lead to the arrest of a former general manager, the general superintendent, and the general storekeeper of the railroad. More members of the gang were later uncovered, with the Illinois Central Railroad plundered out of an estimated minimum of $1.5 million.[43]

Divorce by dictograph hit the news again in May of 1912. Illustrations accompanying the article show a diagram of the device hidden behind a piano in the wife's home while a wire led across the street to where the husband sat with detectives and listened. The husband was Eugene Fallabom of Pittsburgh while the third party (co-respondent) was Jonas McClintock, son of the first mayor of Pittsburgh and one of the city's wealthiest and most prominent citizens. Over the bug, the husband and the detectives heard Mrs. Marguerite Fallabom bad-mouth her husband and McClintock commiserate with her. When he first became suspicious of his wife, Fallabom sought the advice of Edgar R. Ray, the head of the Ray Detective Bureau and before that a New York Police Department detective from 1900 to 1903. Ray suggested the use of a dictograph. A room was rented in the house across the road from Mrs. Fallabom's residence. The transmitter of the bug was hidden behind the piano in a room in the wife's residence. Wires led out the window and up to the roof. Then they ran across the street down the side of the house in which Fallabom had rented the room, through a cellar window and up through the wall inside to the second floor room used by Fallabom. At least two detectives were always stationed in the listening room.[44]

An editorial in praise of the machine appeared in June 1912 and declared, "The attack on the character of the dictagraph seems to have failed. There is one witness that cannot perjure himself. It tells the truth,

A 1912 sketch showing the wiring involved in the dictograph usage that produced evidence in the Fallabom divorce case, and Mrs. Fallabom.

like the scratch of a glacier. Thus science comes nimbly up to the support of justice.... Of course, some people can't believe it." The editor went on to say that the device was doing very well for an infant and "when it is grown it will be made an aid to the publicity bureau and one will be put in the governor's office, the finance and judiciary committee rooms, in the mayor's office, in the political rooms of all parties, so the people will know all about their business, on the inside and outside. What right has public business with any secrets anyhow. Come, let the truth be known, and let the dictagraph, its apostle, be used throughout the land."[45]

Even more publicity came the way of the device when the *Literary Digest* featured the bug in its June 15, 1912, edition, under the headline "scientific eavesdropping." It acknowledged William Burns as the first American to see the "immense possibilities of the instrument in detective work. He is so enamored with it that he always carries one in his pocket. Fictional detectives carry automatics and handcuffs, Burns carries a dictograph." The article went on to list the usual cases such pieces cited, while heaping enthusiasm and praise on the machine. Turner was cited therein as making distinctions between a "commercial dictograph" and a "criminal dictograph," and the "detective dictograph," which, if anything, probably added more confusion. In conclusion the article enthused, "Always listening ... it promises more and more sensational disclosures, more confessions—an 'automatic third degree.'"[46]

Private detective Reed of the William Burns Agency told the investigating committee of the South Carolina Legislature in Augusta, Georgia, on July 17, 1912, that he had talked with a man purporting to be an agent of South Carolina Governor Coleman Blease and that an agreement was made whereby a convict was to be pardoned for $15,000, out of which the governor was to receive $5,000. To support his charges, Reed introduced dictograph records. Those records purported to show that in the New Willard Hotel in Washington, D.C., a man named N. M. Porter (the alias of Reed) pretending to be a Chicago lawyer met with Sam J. Nichols and the pair discussed at length securing a pardon for Gus Deford, a safe cracker who had served a term in the Federal penitentiary in Atlanta and upon his release was taken by the South Carolina authorities and sentenced to ten years for blowing up a safe at Encore Manufacturing Company. At the same time Reed was testifying he told the committee that from messages received from other Burns operatives in Bamberg, South Carolina, that Blease was at that moment closeted with Nichols. The latter man said he had influence with the Governor. That $15,000 was to be divided up with $5,000 to Nichols, $5,000 to Blease, and $5,000 to the Governor's law partner, C. P. Sims. There was nothing in writing. Reed had given Nichols a check for $500, as a deposit. Earlier evidence against Blease came from Burns operatives and Thomas B. Felder, an Atlanta lawyer on a crusade against what he called the governor's grafting. Felder claimed that Blease had received $2,000 for pardoning a convict named Rudolph Babon, around $10 a month from each of the 35 Charleston, South Carolina, "blind tigers" (illegal alcohol outlets) in return for which he gave them the state's protection from prosecution, $5,000 by a railroad line for killing legislation it did not like, and so on. With reference to the pardon deal Reed investigated, over 25,000 words of testimony from the dictograph were placed in evidence. No more was reported on this case.[47]

Through the operatives of the Burns Detective Agency, it was reported from Detroit on July 27, 1912, that 13 Detroit aldermen and the secretary of the Common Council Committee were alleged to have been caught accepting bribes to vote to close a certain street for the Wabash Railroad. The largest bribe was $1,000 and went to Alderman Thomas Glinnan, with $100 bribes going to each of the other 12 aldermen and an alleged $500 to Secretary Eddie Schreiter. All 14 men were said to have been arrested.[48]

Six of the biggest milk dealers in Minneapolis were arrested on October 2, 1912, for illegally combining to raise the price of milk by one cent a quart. County Attorney Robertson got the evidence against the men by planting a dictograph at a meeting held by the dealers a few days earlier. Arrested were Arthur R. Ruhnke, president of the Minneapolis Milk Company; C. A. Nelson, president of Clover Leaf Dairy; W. A. Page, president of Quaker Creamery Company; C. J. E. Johnson and Sam Johnson, officers of the Model Dairy Company; and Oscar Erickson of the Minneapolis Milk Company.[49]

According to an October 27, 1912, news story, James Murdock, who was sentenced by Judge H. Craig of the district court at Newcastle, Wyoming, to be hanged on February 24, 1913, for the murder of John Giachono, was the first man in the United States to be convicted of murder in the first degree through the agency of the dictograph. That evidence was a stenographer's transcription of a conversation heard through a bug, which resulted in the verdict of murder in the first degree. After the killing of Giachono, the prosecuting attorney's office did not have positive evidence of the guilt of Murdock. A dictograph outfit was ordered from New York and so placed that Murdock talking in one room to an accomplice, while he was in jail, was heard in another room. Admissions that he made and were taken down by the stenographer supplied the evidence needed to convict him. The court admitted the evidence over the strenuous objections of the defense counsel, which expected appeal to the Supreme Court. Nothing more was reported.[50]

Operatives of the Burns Detective Agency were reportedly at work with dictographs and stool pigeons in New York in November 1912 on voter fraud. They were said to have run down more than 3,000 cases of false registrations and had reported them to the Honest Ballot Association. That organization was said in the newspaper report to be non-partisan and backed by the likes of the Progressives, the Empire State Democracy, the Republican Club and the Citizens Union. Members of the latter group included Elihu Root, Jr., and John D. Rockefeller, Jr. From the reports that had come to the Honest Ballot Association from the Burns detectives and from 800 watchers (some of them Columbia University students) stationed at the registration places two weeks earlier, the officers of the association thought that in some parts of New York City as many as 2,500 fake votes had been cast in the recent elections. Their evidence was a drop-off in registration in some of the "more notorious" districts, said to be due to the fake voters being scared away and not to a drop in legitimate voters registering. Allusions were made to massive problems such as goons bringing in batches of men from outside areas and then registering them, keeping them around until voting day and then having them, literally, vote early and often. No specific details were provided on how the dictographs were used or on the so-called voter fraud either.[51]

That a force of Burns detectives under the leadership of W. A. Mundell, chief of the Pacific Coast division of the Burns agency, had operated in Nevada during the recent election campaign became known in Reno on November 10, 1912. One of the operatives, named Kelly, was in Senator William A. Massey's office. Massey (1856–1914) was a United States Senator from Nevada from 1912 to 1913. He was appointed to fill the seat after Sen. George S. Nixon's death but was defeated in an election for the remainder of the term by Democrat Key Pittman. Kelly arrived in Reno with a letter of recommendation to Nixon. Finding that Nixon was dead, he sought out his successor who was running to fill the remainder of the late senator's term. He represented himself as a mining man from Australia looking for Nevada properties and in that way established himself in Massey's office where he was wined and dined by numerous public officials, including Nevada Governor Tasker Oddie. Two years earlier it had been asserted that Key Pittman was defeated through the use of money and that this year detectives would be used to prevent that. For two weeks preceding the election the Burns men shadowed the leaders in Senator Massey's fight. On the night before the election Mundell, the chief, outlined the entire plan that had been followed and issued a warning against the use of money by any person. No details were given on the specific use of dictographs, although the article title stated that dictographs were used to prevent fraud.[52]

The introduction of dictograph evidence was the most important event in the trial in Mays Landing, New Jersey, in December 1912 of the nine members of the old Common Council of Atlantic City, New Jersey, on indictments charging them with conspiring to accept bribes in connection with the plan to erect a new concrete boardwalk along the city's waterfront. Burns operatives had been retained to trap councilmen. Those politicians on trial had fought long and hard to exclude dictograph evidence—which was being introduced in the state of New Jersey for the first time—but lost. Witnesses swore the device was in the room where the matter was discussed and Ralph, a stenographer, read from his verbatim transcript. In the entrapment a Burns operative had paid councilmen Malia and Dougherty $500 each and $50 to councilman Phoebus on February 17, 1911.[53]

A dictograph played a part in causing the sudden resignation of the Rev. Alfred G. Mortimer of St. Mark's Episcopal Church, Philadelphia, according to a report late in December 1912. Evidence was said to have been gathered by dictograph and "to have been of such character that it convinced even the warmest friends of Dr. Mortimer." Late in November of 1912 two employees of the parish were dismissed. On leaving they accused Mortimer of grave irregularities. Those charges seemed to carry at least some truth and as a result a bug was placed in the rector's study. When the evidence from the bug reached Bishop Rhinelander he demanded Mortimer's resignation, and got it at once.[54]

A couple of weeks later more details on the Mortimer case became available. The Reverend Mortimer had been thrown out of the Episcopal Church because of "immorality of so gross a nature that decent men would not even mention it, and involving several women and youths," some of whom were members of St. Mark's Church, of which Mortimer had been rector for 11 years. Mortimer was "constantly receiving" women and other visitors in the rectory building—in a small room adjoining his study and sometimes in his bedroom. Several complaints had been made earlier to Bishop Rhinelander. But he could not believe the allegations and insisted on absolute proof. To that end Burns detectives were engaged and laid a careful trap. Dictographs were hidden in the small room off the study and in his bedroom. Those wires went to the basement where Burns operatives were hidden. On the Friday before Christmas 1912 a woman called on the rector and was received in his bedroom. "The unerring dictagraph told the listeners in the basement what passed in that upper room." On the following Sunday a second woman visitor arrived. She was received in the study anteroom—with the same result. Records of those conversations were given to Rhinelander, who confronted Mortimer, got his resignation and gave him five hours to get out of town. He did, at least as far as New Jersey.[55]

A much-heralded event took place in January 1913 when the first practical illustration ever given in the courts of New York City of the accuracy of the device was presented in Judge Malone's court on January 24 in the trial of Mrs. Fannie Dio, a fortune teller. She was charged with attempting to extort $1,000 from Dr. Samuel Tandlich of New York City. New York Police Department detectives Oswald and Hauser told in court how they rigged a dictograph in Tandlich's home and ran wires to the basement, where they listened. Through the bug, they said, they heard a conversation between a lawyer representing Dio and Tandlich, which appeared to reveal a conspiracy to extort money from the doctor. Judge Malone would not allow the detectives to tell what they heard unless it could be shown

The Reverend Mortimer, 1913. The headline of the story explained it all: "Rector who received women visitors in his bedroom is convicted by dictograph and thrown out of the church."

to the jury that the device did really carry conversations. Thus, George Read and John Ambler, New York Police Department "dictograph experts" rigged up the device used by Oswald and Hauser. They placed the transmitter on the back of the judge's chair and strung 50 feet of wire to a jury room outside the courtroom. Oswald and Hauser went to the jury room and Assistant District Attorney Bostwick talked in a conversational tone near the transmitter. The two men heard him 50 feet away through two closed doors and came back into the courtroom and repeated Bostwick's conversation. Then the transmitter was placed in the jury room and Oswald and Hauser stood in the witness stand with receivers to their ears. Court officer Donohue, standing three feet from the transmitter and talking in a conversational tone, directed the men 50 feet away to hold up their hands, scratch their heads, bow, and so on, which they did in concert and correctly. Then Judge Malone directed the evidence of Oswald and Hauser to be admitted as evidence.[56]

In the Dio court case Tandlich alleged Dio had, in October 1912, written nine letters to him at his home in each of which she demanded $1,000. In another letter the threat was made that unless he paid, his professional reputation would be harmed. A police shorthand expert was assigned to the case, a bug was installed in the Tandlich home and the doctor sent word to Dio he was ready to arrange a settlement. She sent the lawyer to see him and the case was based on evidence from that conversation.[57]

Kelly Turner, inventor of the dictograph, was a witness in the Dio case. He was there to attest to the efficiency of his device. He, wrote a reporter, "had the utmost confidence in his instrument, which he perfected in 1907 after devoting the best years of his life to it." Repeated was the idea that Turner never sold any of his devices because his conscience forbade him to allow the instrument to get into the hands of persons not known to him "as law abiding and trustworthy." He worried, "There is no limit to the possibility of the uses of the dictograph for criminal purposes. Hijackers could reap a harvest with it, trade and business secrets could be stolen, the hiding places of valuables revealed, the plans of police and peace officers overheard." So he only leased his machines. The device exhibited in court was identified as a dictograph made for detective purposes. According to the journalist, "Mr. Turner said that he had made and leased several hundreds of the instruments." Turner added, "My instructions in my office are to make it as difficult for a man to lease one of these instruments as it is to cash a $10,000 check in a bank where he is not known." He claimed that in a room 15 feet square his device would record every sound, even a whisper. As a last thought Turner emphasized, "I have made and leased several hundred of them, but in every case I know exactly for what purpose the instrument is used, and we never lease one unless we are sure it is for legal and proper purposes. None is ever sold outright."[58]

Early on the morning of February 1, 1913, the jury in Judge Malone's trial of Dio returned a verdict of guilty against the woman. Malone sentenced her to Auburn State Prison for not less than three years or more than five years and six months. Dio had alleged Tandlich performed an "illegal operation" on her sister and hence the demand for hush money. For the first time in the history of the criminal courts of New York State a defendant was convicted solely on dictograph evidence, it was said.[59]

Charged with having received large bribes to vote for Col. William Seymour Edwards for U.S. senator, four members of the House of Delegates and one state senator were arrested on February 11, 1913, in Charleston, West Virginia, by G. B. Bentley of the Burns Detective Agency, who posed as an agent of Edwards. Bentley paid $20,000 to the five men and then took them one by one into another room and turned them over to the prosecuting attorney, the sheriff and several deputies. A dictograph played a part in the arrest of the five Republicans. One arrested was Delegate Rath Duff while another was Delegate S. U. G. Rhodes. Rhodes placed the name of Isaac T. Mann for U.S. senator before the Joint Assembly, but on February 11 he voted for Edwards. William J. Burns was in Charleston on February 11 to witness the arrest of the legislators and said that at the start of the investigation they found it difficult to obtain evidence as the politicians refused to talk very much, preferring to resort to the use of pad and pencil, using their fingers to indicate how much they wanted, and so forth. "In a short time, however, we were able to obtain the evidence through the dictographs ... installed in their rooms."[60]

Four dictographs reportedly were found in the Pennsylvania House of Representatives by one of the officers of that body in March 1913. He had torn them out and said he did not know where the wires went. A week earlier one of the legislators said that he had positive information that eight devices had been installed in the House side and four in the Senate. He did not know locations. Investigation then failed to disclose that any dictographs had been acquired by the state. The ones found were under the seats of four members and were so distributed that they would record conversations held by certain groups of members. There were small ventilators at the bottom of each desk and the bugs were tucked up in them. "There were grounds for suspecting that the Burns detective agency is responsible," wrote a reporter, although no details were offered. It was also said that dictograph searching parties were then being organized.[61]

An editorial on the above situation noted, and emphasized, that no politician was safe, even in what he thought was a private place. "It might seem, after all, as if this ingenious instrument might prove an important adjunct of the machinery of virtue, and it may even enforce the rule of morals that honesty is the best policy when dictographs are around."[62]

After two years of work by a detective agency, in the course of which the dictograph played a major role, Joseph Moriarty, alias William J. Leehan, formerly employed in the home of Mrs. E. M. Korne of Pittsburgh, was in jail as of April 4, 1913, charged with the murder of Mrs. Charles Turner of Lakewood, New Jersey, in April 1911. Leehan was lured from New York to Fort Lee, New Jersey, and arrested by detectives whom he thought his intimate friends. Turner was murdered outdoors on a path and no definite evidence was discovered but clues were said to have pointed to Leehan, who was reported to have quarreled with his wife on the afternoon of Turner's death. He had been seen entering the same lane as Turner at about the same time. Leehan and his family disappeared from Lakewood a few weeks later. His disappearance reawakened interest and detectives finally found Leehan in jail in Newark for wife beating. He was trailed to White Plains, New York, and a dictograph was installed in his apartment and connected with rooms next door, where detectives listened for weeks. Leehan moved several times to different parts of White Plains but investigators always found a way to install the bug at his new locations. After enough evidence had been gathered through the use of the bug, Leehan was arrested.[63]

Lacking the ability to record conversations automatically had always been a serious drawback for the dictograph and research to remedy that situation continued off and on but was never successful, in the period covered by this book. Late in April 1913, a news story heralded a supposed breakthrough in that area. "A self-recording dictograph, which could not only overhear a conversation in a room where its presence was not suspected, but could make a full record of the conversation, whispers and all, on a phonograph cylinder located some distance away, is being exhibited by K. M. Turner, the inventor of the dictograph at his office in West 42nd Street," wrote the journalist. Such a device had been much sought-after, he continued, ever since the dictograph was placed on the market some years earlier. It was acknowledged that in court cases up to then the Burns Detective Agency had been forced to submit stenographic notes, the authenticity of which it had to prove. For years Turner had worked to overcome the problem. Even then a businessman could dictate his letters to be mechanically recorded on cylinders. In detective work Turner said that the absence of the self-recording feature had proven "an almost insurmountable difficulty." It had been necessary to make the instrument so that two detectives could listen instead of one. In some court cases the dictograph's evidence had been thrown out because a single detective's transcription was thought not to be reliable enough for a conviction. At his demonstration, Turner admitted that when four reporters conversed normally the conversation was captured clearly and faithfully, but when talk fell to a whisper it was not recorded properly. However, Turner argued that a detective would always be present in the listening room and he could fill in with his stenographic notes for the faint or missing parts from the recording device. His device did not work and quickly disappeared from sight.[64]

William Burns and the operatives with his agency were the people who used the dictograph the most and who got, by far, the most publicity for using the bug. But at the start of May 1913 it was reported that Kelly Turner and William Burns had parted company. Turner had called in all the dictographs leased by Burns and in use by his operatives in the field. That action was the culmination of a year's "misunderstanding," said a reporter. Burns, it was said, tried to dictate who should be permitted to use the bug and objected to paying the same rental rates for the device as were charged to other detective agencies. He argued he was entitled to a special price and other concessions that would give him an advantage over other detective agencies. Turner said that in view of the prosperity that came to the Burns agency in the wake of using the dictograph, he could not agree with Burns's suggestions. They reportedly argued regularly over money. In view of the breach, which he recognized as growing, Burns sought out other devices to take the place of the dictograph. He tried the telegraphone, which recorded the voice on a wire spool. However, this recorded so poorly it was of no value and Burns gave it up. Then Burns called at Menlo Park to try to and interest Thomas Edison in perfecting a device, but Edison was involved in an important problem and would not even see Burns. Then an open break took place and Turner issued his recall order. In the end, Turner wanted more money from Burns for the use of the devices as they brought so much money to Burns and his agency. Said Turner, "Burns was to make whatever charges he felt proper to his clients for the use of this instrument in their cases and to get the benefit of the enormous increases in profitable business it brought to his agency, while we were to get practically nothing but sweet praise." Burns refused to give any more money to Turner and even wanted a reduction in the leasing rates. He felt Turner should be content with the extra publicity Burns brought to the device, which, presumably, would lead to more rentals overall.[65]

A photograph showing an unidentified detective from the Burns Detective Agency at the listening post with a dictograph. This Library of Congress photograph is undated but it would have been 1910–1913.

A news account at the end of April 1913 observed that operatives for the Burns agency had reportedly trapped officials of Ocean City, New Jersey, and other state officials in the act of seeking bribes. Those operatives represented themselves as promoters of a trolley line along the New Jersey Ocean Boulevard. Arrest warrants were reportedly in the works for several state politicians. Methods used by the operatives were nearly the same as those they had used to trap Atlantic City, New Jersey, grafters who were willing to sell their votes for a mythical concrete boardwalk. Three Burns detectives and one woman rented a suite of rooms at a hotel and then bugged the place and laid wires under the floor. This article was different in that it was the first to mention that the Burns agents used the detectaphone, a new apparatus similar to the dictograph. Obviously, Burns had found someone with a slightly different device. It must have been different enough to avoid patent litigation, as they seemed to be no lawsuit over patent infringement by Turner against Burns. This detectaphone worked exactly the same as Turner's device; it did the same things in the same way, and it could not record automatically. According to one account, the inventor of the detectaphone was Chicago scientist and inventor Montraville Wood.[66]

In Pawhuska, Oklahoma, on May 21, 1913, County Judge Charles E. King, Sheriff E. A. Willison, and Under-Sheriff E. E. Sams were arrested and charged with bribery on a complaint issued by County Attorney C. M. Cope. Evidence secured through the use of the dictograph in Cope's office furnished the basis for accusations of a wholesale plot to extort money from Osage County bootleggers. The three officials came to Cope and Assistant County Attorney Ottamar Hamele and tried to get them to join in the extortion ring. Those two encouraged the three to explain and elaborate, which they did—to a hidden dictograph. A system of extortion was explained that was designed to extort about $100 a month from each bootlegger in Osage County. The three men arrested had all been sworn into office in January 1913 and were all Republicans. The other two were Democrats.[67]

A brief report in June 1913 observed that the nationwide use of the dictograph in criminal detection was under discussion at a session of the convention of the International Association of Chiefs of Police that was then underway. The resolution committee was reportedly prepared to introduce a resolution favoring the adoption of that device, noting that it had been used successfully by Burns operatives in the McNamara dynamite case.[68]

On July 12, 1913, it was reported that a strike was to be ordered in the next day or two on the 42 Eastern railroad lines, whose officials had refused demands for a wage increase; some 100,000 employees were involved. A meeting of union executives had been called to order that morning at the Broadway Hotel, in New York City, as planned and scheduled. However, in order to avoid possible dictograph plants the whole convention was immediately ordered to move to another venue in the city. Only two of the union executives (out of some 900) knew of the switch. Said one of the union executives on the committee, "This precaution was born of a bitter experience. On the occasion of two different meetings in Chicago we learned that dictagraphs had been planted in the halls used, and that our every remark had been recorded. It happens, however, that we captured the dictagraphs and have them yet."[69]

Two men who were accused of having tried to defraud a burglary insurance company and to bribe an adjuster employed by the company to put their claim through were arrested on August 12, 1913, in New York City, after accepting a roll of marked bills from the adjuster and talking about the affair for some time while a dictograph carried the entire conversation to a police stenographer sitting in the next room. Nathan Maier, 24, and Harry Baltimore, 39, were the men arrested. Maier had reported a false burglary at his home, a supposed $700 loss of jewels. He then recruited Baltimore, an insurance broker, to help in the scheme. The claim was investigated by Gerard Luisi, who met the men and told them he agreed with police that their claim was "fishy." They admitted to Luisi the claim was false and they offered to cut him in on it. He reported this to his company, with the result that the police were called in and the set-up with the dictograph was arranged. A meeting was arranged in a room at the Hotel Seville with a police detective and two stenographers stationed at the listening post in a second hotel room.[70]

District Attorney W. W. Bridgers of El Paso, Texas was involved in the Orner case in Van Horn, Texas, and in the Godsey case in El Paso. Information revealed in October 1913, with respect to the Orner case, that "the district attorney's office in El Paso has been using a dictagraph on various lawyers in El Paso and that one was used in connection with the witness Maese in the Orner case." The story continued, "The members of the grand jury were treated to their first experience, it developed, with the dictagraph when they were escorted to a room in a local hotel this week, where one end of the dictagraph was installed, and each listened intently to what was going on in a room some distance off."[71]

Conversations between former New York State Senator Stephen J. Stilwell and some of his visitors were caught at Sing Sing Penitentiary in the summer of 1913 by detectaphone. That device was installed in Sing Sing by Burns Agency operatives on July 25. The receiver was placed in

Warden James Clancy's parlor. From the parlor the wires went up through the ceiling to a room above where the stenographers and listeners were quartered. Nine conversations were heard through the bug, the first on July 30 and the last on August 30. Only three people knew of the installation: the stenographer, Warden Clancy, and investigator John A. Hennessy. In May 1913 Stilwell had been convicted of accepting a bribe and automatically forfeited his seat in the New York State Senate. He was sentenced to a prison term of not less than four years and not more than eight years. He began to serve his sentence at Sing Sing on July 15, 1913.[72]

A week after the above report, a photo of William Burns was published in newspapers around the nation. The text explained that he was being guarded by nearly 100 of his detectives and that his family had received reports that $10,000 had been raised to buy his death and forestall detectaphone disclosures involving high-level politicians in New York State. A reporter remarked that wherever "the world famous detective goes his men, heavily armed, keep him in sight."[73]

State officials in San Francisco used a dictograph concealed in a letter file on a bare desk to secure the evidence of bribery against W. S. Card, a quack specialist, who was arrested on November 26, 1913, in San Francisco. After frustrating several attempts to get a witness, Card was finally lured to an office of the State Board of Medical Examiners to keep an engagement with A. G. McDonnell of the board. Two floors below sat a stenographer, a Burns detective and Dr. C. B. Pinkham, secretary of the State Board of Medical Examiners.[74]

Guy Downing, alias James Murray, a brakeman for the Northern Pacific Railroad, had a scheme to defraud the railroad out of $750 with a fake accident claim. However, it was uncovered through the use of a dictograph. Downing was declared by officials to be a "professional personal injury shark." He had secured work as a brakeman and in December, shortly after he had gone to work, fallen from a freight car, claimed to be seriously injured, hobbled around on crutches, and so on. A company physician could find no trace of injury, yet Downing demanded $750 in compensation. A dictograph was placed in the downtown hotel room where Downing resided and special agents of the railroad listened to the other end of the wire. One night Downing and a friend were conversing in the room and Downing told his friend of his scheme to bilk the railroad.[75] He broke down on January 24, 1914, when he was grilled by Deputy Prosecutor Askren and admitted he attempted to extort. He was charged with grand larceny immediately, taken before Judge Clifford, and sentenced to a term of from one to 15 years in the Walla Walla penitentiary.

Kelly Turner was not much heard from but he occasionally did pub-

A 1913 diagram showing how the bugging of Dr. Card was carried out. Shown is private detective McDonnell.

licity work. On the evening of February 21, 1914, he demonstrated the device he invented at the National Press Club in Washington, D.C. Turner had arranged connections for all the rooms of the club with his wires and demonstrated his device.[76]

Roy Rutledge and a Mr. Grider were arrested in Henderson, Kentucky, on February 25, 1914, in connection with counterfeit bank notes. Police knew that if a dictograph was to be effectively used the men would have to be together. On a Friday afternoon Jailer Howard at Owensboro, Kentucky, was instructed to take Grider out of his cell, ostensibly to take a bath. While the jailer remained with him the government agents were admitted to Grider's cell and planted a dictograph under his cot. The wire ran into the outer jail office. After his bath Grider was asked if he wanted to see his friend Rutledge. He said yes. Rutledge was taken to his friend's cell where the two men quickly incriminated themselves with their conversation.[77]

It became known on January 11, 1914, in Houghton, Michigan, that a dictograph was planted in the law headquarters of the Western Federation of Miners for five weeks in the summer of 1913, when the seat of trouble was at Calumet, Michigan. The labor leaders and their attorneys held practically all of their conferences at the beginning of the strike in the very office where the dictograph was hidden and a complete report was obtained by Burns operatives. A second dictograph, it was said, was planted in the Hancock, Michigan headquarters of the union in the Scott Hotel. The Burns men had originally been brought into the strike district by the mine operators. Their reports were said to have also been turned over to Special Prosecutor George Nichols, who had charge of the grand jury that was investigating both the deportation of union president Mayer and the conspiracy that had been laid at the feet of the labor chiefs.[78]

According to a March 8, 1914, report in a New York City newspaper, the detectaphone was then being used frequently in the executive offices of big corporations to find out what the chief employees were doing. That fact was disclosed in the Bar Association rooms of the creditors of the Palmer & Singer Manufacturing Company, makers of Palmer-Singer automobiles. The bug was put in the office of Charles A. Singer, former president of the concern, who had been transferred to a lower executive position in the spring of 1913 when John T. Pratt, a member of the wealthy Standard Oil family, bought an interest in the company and installed his representative, Clyde D. Knapp, as president. Singer was told of the planted device by a discharged employee and on investigation found the wires connected to Knapp's office. At least 12 such devices had been planted in the offices of other people in that firm. Singer added that he had been told that dictographs were then employed in many of the offices of large corporations.[79]

In Portland, Oregon, evidence from a dictograph was introduced at the breach of promise suit brought by Mrs. Gertrude Gerlinger (a divorcée) against Lloyd Frank, a prominent merchant of that city. At the conclusion of that trial Gerlinger was awarded $1, a token sum. So, on April 4, 1914, she filed suit against Frank's attorneys, the William Burns Detective Agency, J. D. Huddleston (proprietor of the Buena Vista apartments), and janitor F. E. Glenn, for $50,000 on grounds of trespass. She alleged the attorneys maliciously employed the Burns agency to place the detectaphone in her apartment with the assistance of the other two men named in the suit and through that device conversations of a private nature were transcribed and later published to the community when such reports were introduced at the trial.[80]

Investigations into the mysterious death by shooting of Mrs. Louise Bailey in the office of Dr. Edwin Carman on June 30, 1914, in Freeport, New York, led to the discovery that a dictograph had been installed in the physician's office with a concealed wire leading upstairs to the bedroom of Mrs. Florence Carman. Only a few days earlier Mrs. Carman had obtained additional batteries so the device would transmit more efficiently. The story of Mrs. Carman's negotiations for the installation of the device was told by Gaston Boissonnault, manager of the dictograph department of the General Acoustic Company of New York. He said she came to his office on May 19 of that year and, after some verbal fencing, admitted to wanting a device installed so she could overhear the conversations of her husband with his women patients. She said, "I don't want to get a divorce. I simply want to overhear what he says to women. I have a young child whose name I must protect." She went on to explain that she believed her husband was too friendly with a woman he sometimes employed as a nurse and that she had seen him kissing that woman. Dr. Carman said the night of the murder was the first time he had ever seen Bailey, that she came to see him to get a prescription for malaria medication.[81]

In the Bailey murder case the bullet that killed the woman had been fired through a window of Carman's office. Mrs. Carman was called to view the body but she said she had never seen her before. Boissonnault added that he had explained to her that she could neither buy nor rent the dictograph if it was to be used in a divorce action. On June 23 she came to the office and said her husband did not go out of town much but that she could arrange a trip to New Jersey for them and during that time the firm could install the bug. She came to Boissonnault's office after the installation to pay $50. On the stand, Boissonnault admitted there were three dictographs hidden in his own office and he did not know if any of his employees were listening to the talks he had there with Florence Carman.[82]

Florence Carman went on to be tried twice for the murder of Bailey. The first trial ended in a hung jury. She was acquitted at the second trial. And then it was all over. The murderer of Louise Bailey was never found.[83]

Another article on the general pervasiveness of the dictograph appeared in print in November 1914. It claimed that widespread eavesdropping had become a common thing and was also being used to save time in commercial houses. In New York City it was said that 50 dictographs were purchased every week by jealous husbands and wives but the number of devices used for domestic disputes was small compared to the number being installed for commercial purposes in offices, stores and factories. Reportedly, agents of many corporations found the device to be of service in detecting fraudulent claims for damages. Each claimant and his witnesses were made to wait a few minutes in a "very completely dictographed" room. If a claim was fraudulent, those people in that bugged room were likely to take that waiting time to rehearse their stories and agree on the testimony they were going to give. This article was vague, with little or no numbers, details, specifics, and so forth.[84]

A member of the West Virginia Legislature in Charleston proposed, in January 1915, to introduce a measure to forbid the use of dictographs, except where the person on whom it was used gave his permission and consent. Nothing came of the proposal.[85]

In Atlantic City, New Jersey, State Supreme Court Justice Kalisch set aside the conviction of two men found guilty in a lower court in that state on graft charges on the evidence of a Burns detective and his dictograph. The operative's testimony was supported by notes of a supposed conversation that a stenographer stated came to him over the bug. The court ruled that this was not sufficient corroboration because the stenographer could not see either of the accused men and had never heard either of them speak before his stint in the listening room. A new trial was ordered.[86]

An editorial on the issue of privacy and the dictograph appeared in several newspapers in America in May 1916, originating in a Philadelphia paper. As far as the editor was concerned, there was a question as to whether a man's home was his castle: whether another man, could tap his phone, install a dictograph, and so on with impunity. He acknowledged the fact that such spying was a necessary means of hunting down crimes and criminals and "that it is essential that the police shall have a free hand for the tasks assigned them, but it is equally obvious that if it is to be the recognized privilege of every detective agency or individual sleuth—or even the police without the formality of a special warrant—to tap a telephone or secretly install a detectaphone, what have been considered sacred rights of the individual are abandoned."[87]

When a surface car strike took place in New York City in the summer of 1916 the union organizers held their strategy sessions at the Continental Hotel. On August 12 a woman moved into Room 742, which was exactly over the room used as a secret council chamber by the Carmen. As she walked across the room, her foot sank in to a depression in the floor. Inspection showed a square had been cut out of the carpet. Underneath, a hole big enough for wires to press through had been cut through the floor to the plaster of the room underneath. Hotel electricians and the house detective agreed it was a place where a dictograph had been planted. Two men had occupied the room during the deliberations of the strike committee in the room beneath.[88]

Eight members of an alleged gang of 60 were arrested in Chicago on September 18, 1916, on charges of having fleeced wealthy men and women out of more than $250,000. Five men and three women were arrested. The schemes were designed to "compromise" victims and then blackmail then later. Federal agents made the raid and the arrests. According to the story, "The raid was made only after the apartment building had been [sic] literally sown with dictographs. These were connected with an adjoining building where the detectives hid."[89]

In Bremerton, Washington, the discovery of a secretly installed dictograph in March 1917 in the office of City Attorney Marion Garland, with the receiving end in the basement of the hardware store owned by Mayor M. E. Giles, led to accusations, talk of conspiracies, and so on. Garland and his law partner, Arthur C. McLane, discovered the bug in their office. Sheriff Dan J. Davis traced it to the hardware store. They allegedly held up the Mayor at revolver-point and removed the device. They arrested L. L. Williams, a Seattle attorney who was said to have been listening in on the machine at the time of the raid. It was reported that the "formality" of a search warrant was dispensed with in entering Giles's store. Deputy Sheriff Walter Thompson (who was with Davis) was under arrest on the complaint of Mayor Giles for trespass. Rumor that the four men named had been in a conspiracy to permit the operation of illegal gambling and drinking spots and other shady deals abounded. Giles opposed them. On the other hand, those four claimed Giles was trying to get information so he could tip off those subject to raids proposed by the sheriff and the city attorney. Reportedly, the bug had been in place since February.[90]

Anthony Senes, head of the Senes Detective Agency, Harry Van Pelt, another private detective, and Jack Jacobson, an apartment building janitor, were all charged with eavesdropping in having installed a dictograph in the home of Nathan Newman, an antique dealer who lived in New York City. All three were held for trial in Yorkville Court. It was a mystery as to why the dictograph had been installed in the Newman apartment. Two private detectives posed as electricians who were in the building to fix the wires. The building janitor assisted the two sleuths in their operations. The defense argued that the law of eavesdropping was obsolete; that it was in place before dictographs were invented and that it had to do with listening at keyholes. Prosecuting the case was Ferdinand Pecora, Assistant District Attorney. He contended that the law against eavesdropping made it a misdemeanor punishable by a year's imprisonment or a $500 fine, or both, to listen through a keyhole. He asserted that a dictograph was in effect an "elongated ear." The defense cited the eavesdropping law to show that dictographs were not mentioned in that law: "A person who secretly loiters about a building, with the intent to hear discourses therein, and to repeat or publish the same to vex, or annoy or injure others, is guilty of a misdemeanor." Pecora argued that the law did not permit the introduction as evidence of conversations between man and wife, because that would violate the sanctity of the home. "Surely it would not look with favor upon the evidence of a detective listening through a dictograph." Earlier, defense attorney Koelble had argued, "They were not eavesdropping in the sense of looking for scandal. As duly licensed detectives they had a right to follow their legitimate business. If they were searching for evidence of gambling their evidence was credible. If they were looking for evidence in a divorce case they ought not to be condemned." Magistrate Robert C. Ten Eyck ruled, "The defendants overheard by dictagraph. This section does not refer specifically either to listening by keyhole or by dictagraph. They were eavesdropping in my opinion and I hold then in $500 bail each for trial in Special Sessions." No outcome of the case was reported.[91]

The Newman case provoked a letter to the editor from C. H. Lehman, president of Dictograph Products Corporation of New York City. That letter, dated August 28, 1919, was an open letter to the District Attorney and to the New York Police Commissioner with regard to the "misuse" of dictographs. He said the Newman case brought up the problem of regulating the device and other secret listening devices to the public authorities of the city, state, and national governments "and to its definite legitimate use on the part of reputable organizations." Lehman added that during World War I, his company voluntarily discontinued the sale of all dictograph listening devices except to the War Department and the various other secret service organizations and that for more than two years the firm had refused to sell dictographs to private detective agencies and to individuals unless they could assure his firm that they had "a definite legitimate reason for its use." He then pointed out that instruments used in the Newman case were not genuine dictographs or obtained from his company and "the opportunity given to irresponsible and possible blackmailing agencies though the use of such instruments is a public danger and should be done away with."[92]

William J. Flynn, chief of the Investigation Division of the U.S. Department of Justice, discovered in October 1919 why New York City had been a wide-open booze town in spite of Prohibition. According to Flynn, the reason was a clique of Department of Justice agents working with a former agent and others to protect every restaurateur and saloon keeper who would pay them graft. In September Flynn had wondered why it was possible for almost anybody to drop into almost any bar or restaurant and get all the hard liquor he wanted—usually served in a tea cup or disguised as ginger ale. Flynn furnished a squad of men to investigate, after getting approval and permission from District Attorney Caffey. Of that squad of investigators, two would be arrested. It took only a week to discover the location of the graft ring and the names of many of the members. Flynn kept everything under observation for a month before breaking up the gang. His agents had a series of dictographs planted and also tapped the telephone lines of those involved. Headquarters of the gang was at New York City's Hotel Commodore.[93]

E. A. Burdick, chief of police at Hammonton, New Jersey, and friend of Charles White, a murder suspect then in his custody, came forward in December 1919 to declare that county officials employed a dictograph to eavesdrop on the conversation of relatives and friends with White and his co-defendant, Mrs. Edith Jones. The visitors were admitted one at a time to a room in a wing of the jail to see White and their conversations recorded; then Mrs. Jones was brought in and the performance was repeated. Burdick pointed out that under normal circumstances the visitors would have been admitted in a body to see White and Mrs. Jones.[94] That led to the whole three-man police force of Hammonton being dismissed shortly after coming under verbal attack from County Detective J. P. Wilson, for failing to help him. It was another development in the case, the killing of a man named Billy Dansey. When asked why the men were dismissed, Mayor Boyer said they had taken part in a street fight a few nights earlier. That dictograph had been hidden behind a calendar on the wall of the room in which first White and then Jones were taken to converse with visitors.[95]

A temporary injunction restraining Burns Agency operative H. R. Erlbrook, Seth B. Orndorff (sheriff), and J. H. Rogers (U.S. Marshal) from installing a dictograph in the cell occupied by Carlos Helmus at the county jail in El Paso, Texas, was granted by Judge Duval West of the United States Court at San Antonio in May 1920. Helmus was charged with embezzlement from the First National Bank of El Paso. Helmus applied for the injunction following the alleged discovery of a dictograph in his jail cell. Nothing more was reported on this case.[96]

In the fall of 1921 Alice Clement, one of the first policewomen of that city, remarked that Chicago mothers whose daughters were being courted often used a dictograph as a chaperon. The machine was hidden in the living room, explained Clement, and kept mother informed. She may or may not have been speaking tongue-in-cheek. In any case, the dictograph had moved into popular culture during its first decade. References to it could be found in stage plays and movies.[97]

A slightly earlier stage play dealt with wiretapping. *The Only Law*, written by Wilson Mizner and George Bronson Howard, was produced by Walter Lawrence. It was a play about the New York City Tenderloin. The hero was a professional wiretapper who carried his equipment around in a suitcase and kept just one step ahead of the police. Produced in 1909, it went on to play all over America. The story had Jean, a young chorus girl, in love with MacAvoy, a young man of no principle who took advantage of her love for him to borrow money off her while he lived in idleness. A friend of the two was Spider, a crook who believed the world owed him a living and proceeded to collect it by fleecing those with more money than he had by means of the wiretapping game.[98]

An ad for a local cinema in Bismarck, North Dakota, on August 1, 1912, had four features on the bill. The second one listed was called *Exposed by the Dictograph* and was described as "An intense drama showing for the first time in motion pictures the world famous dictograph."[99]

Never one to be shy of publicity, William Burns wrote and starred in *The Exposure of the Land Swindlers,* a 1913 release by the Kalem Company. It was hyped as the story "of one of the greatest achievements in bringing criminals to justice." A contemporary news article states that one scene "faithfully reproduced" the hall of the House of Representatives in Washington, D.C. Most of the article was given over to Burns to tell the story of the land grafting case upon which the film was based. That film was ready to be released just after the time that Burns and Turner had their split. The film prominently featured the dictograph and Burns tried to get Kalem to remove the device from the film (said to have cost $15,000 to produce). Kalem did not want to spend the extra money and said no, but did make some changes reducing the visual impact, and so forth, of the dictograph in the film. Turner went to court to stop that and forced Kalem to release the film as originally made.[100]

A bit of unsigned doggerel appeared in various newspapers at the end of December 1912: "Maud Miller on a summer's night/Turned down the only parlor light./The judge, beside her, whispered things/Of wedding bells and diamond rings./He spoke his love in burning phrase/And acted foolish forty ways./When he had gone Maud gave a laugh/And then turned off the dictograph."[101]

A new mystery play appeared at the Criterion Theatre in New York City in January 1913. It was called *The Argyle Case* and was written in collaboration by William J. Burns, Harriet Ford and Harvey J. O'Higgins. In

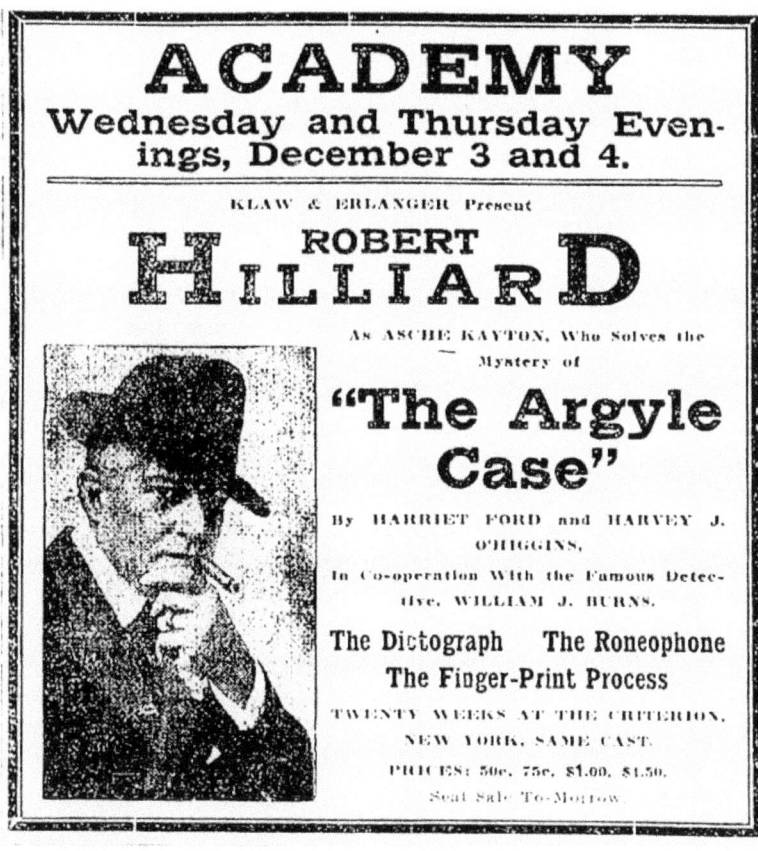

A 1913 ad for a stage play *The Argyle Case,* which also had much involvement from William Burns. Note that the text mentions the "Roneophone." This was a device that supposedly, when attached to a dictograph, automatically recorded the conversation. It did not work and was already dead as an invention by the time this play surfaced.

the play actor Robert Hilliard said, to a murderer, "Did you ever see a dictograph? Don't be afraid—it won't bite you. It has the largest ears in the world and it makes a fool look like a jackass." Reportedly the play was successful. An illustration in a newspaper showed the Hilliard character using a dictograph with the newest attachment, the roneophone, which recorded the conversations automatically. In the play the detective character played his own words back to the murderer who, of course, did not know the room was bugged. The problem was that the roneophone—one of many attempts to automatically record from the bug—had never worked and no such device then existed.[102]

By 1920 the groundwork for the modern surveillance state had been established. All citizens were potential targets, although a good deal of effort was needed to accomplish a bug. Each target had to have devices attached either in the home or office, or on wires leading to that home or office. Listening posts had to be established and one or more people had to be at those posts in real time. It was expensive and time-consuming. The last necessary piece for "perfect" surveillance had not then been invented—the automatic and mechanical recording of overheard conversations by machine. But, of course, it would come. By 1920 such activity was illegal everywhere in America. Many cases described in this book featured bugging by federal, state, or local authorities and it appeared that in no case did any such authority ever seek, or receive, approval for such activity from an established, accepted judicial review process. That is, the police and other authorities simply went out and bugged whomever they wanted whenever they wanted. The beginnings of cooperation between state officials, police departments, and private corporation had already begun. That tradition that was established in this era of state surveillance and cooperation between public and private eavesdroppers was rarely if ever questioned or even mentioned. And, of course, it was a tradition that apparently continues to this day. Every advance in technology has made it easier to perform surveillance on the population until we have reached the stage where the state and the corporations have, with respect to online activities, complete records on all citizens, not just with respect to conversations, but also with respect to where they are much of the time. Yet there is little opposition. George Orwell's protagonist in *Nineteen Eighty-Four*, Winston Smith, was under a similar constant surveillance from the telescreens, as was all the population other than the elites in that novel. Before he "accepted" the system he had to be both physically and mentally broken over a considerable period of time before he would utter the words "I love Big Brother" and really mean them. Today, though, people have not just accepted those telescreens (Facebook and other so-called social media sites), but embraced them. That "Look at me" universe has made things easier for the eavesdropper.

Chapter Notes

Chapter 1

1. *Smoky Hill and Republican Union* (Junction City, KS). September 6, 1862.
2. "Latest news." *Edgefield* (SC) *Advertiser*. September 16, 1863.
3. "Telegraphic secrets." *New Ulm* (MN) *Weekly Review*. October 25, 1862.
4. "Wire-tapping." *New Ulm* (MN) *Weekly Review*. April 2, 1884.
5. "The field telegraph." *Wichita Eagle*. June 12, 1887.
6. "A boy spy in Dixie." *National Tribune* (Washington, DC), May 31, 1888.
7. "Morgan's raid." *Wichita Eagle*. July 29, 1888.
8. George L. Kilmer. "Bold John Morgan." *Princeton* (MN) *Union*. January 2, 1896.

Chapter 2

1. "Stealing electricity." *Sun* (New York) July 20, 1887.
2. *Evening Bulletin* (Maysville, KY). August 1, 1892; *Las Vegas Free Press*. July 14, 1892.
3. "Tapped the wires." *San Francisco Call*. June 29, 1895.
4. "A power that cannot be stolen." *Forest Republican* (Tionesta, PA). March 17, 1897.
5. "Can electricity be stolen." *Sacramento Record-Union*, May 5, 1897.
6. *Suburban Citizen* (Washington, DC), May 26, 1900.
7. "Jersey lawmakers adjourn." *Sun* (New York). April 1, 1897.
8. "Electricity is property." *Evening Times* (Washington, DC). April 29, 1897.
9. "Electric wire tapped." *Scranton Tribune*. January 7, 1898.
10. "Alleged electricity thief." *St. Paul Globe*. July 21, 1898.
11. "Systematic methods." *St. Paul Globe*. September 8, 1900.
12. "Electric light wire is tapped." *San Francisco Call*. December 11, 1901.
13. "Finds druggist Beck is guilty." *San Francisco Call*. January 3, 1902.
14. "Fined for tapping the electric wires." *San Francisco Call*. January 5, 1902.
15. "Claim he stole light current." *San Francisco Call*. March 28, 1903.
16. "Ferry Cafe. 16 Market Street." *San Francisco Call*. April 19, 1903.
17. "For stealing electricity." *Salt Lake Herald*. December 29, 1905; "W. T. Conway is convicted." *Salt Lake Herald*. January 4, 1906.

18. "Puts a thief on wire." *Salt Lake Herald*. October 27, 1909.
19. "South Omaha man fined for tapping light wire." *Omaha Daily Bee*. August 20, 1911.
20. "Held for tapping Edison Co.'s wires." *New York Tribune*. August 16, 1913.

Chapter 3

1. "Sound sent by wire." *New York Tribune*. March 31, 1877.
2. "Is the telephone a failure?" *Stark County Democrat* (Canton, OH). January 17, 1878.
3. "General brevities." *Iola* (KS) *Register*. January 3, 1879.
4. "Pranks of telephones." *Princeton* (MN) *Union*. July 8, 1880.
5. "Message in fragments." *Sun* (New York). March 23, 1883.
6. "The printing telegraph." *Richmond* (VA) *Dispatch*. April 29, 1886.
7. "The field of electricity." *Omaha Daily Bee*. February 24, 1898.
8. "A revolution in telegraphy." *Evening Times* (Washington, DC). December 27, 1898.
9. "Wireless telegraphic codes." *New York Tribune*. October 10, 1907.
10. "Wants to protect telephone talks." *New York Times*. March 22, 1914.

Chapter 4

1. "Tapping news conduit." *Sun* (New York). July 8, 1883.
2. Ibid.
3. "Tapping the wires." *Daily Dispatch* (Richmond VA). July 8, 1883.
4. "Pirates of the wires." *Juniata Sentinel and Republican* (Mifflintown, PA). July 25, 1883.
5. Ibid.
6. "Pool sellers swindled." *New York Tribune*. October 14, 1883.
7. "Not caught." *Salt Lake Herald*. October 16, 1883.
8. "Swindling by telegraph." *New York Tribune*. October 17, 1883.
9. "Tapping the poolsellers' wire." *Sun* (New York). October 18, 1883; "The pool mystery discovered." *Los Angeles Herald*. October 19, 1883.
10. "Badly cinched." *Sacramento Record-Union*. October 19, 1888.
11. "The wire was tapped." *Washington Post*. October 19, 1888.
12. "Tapping the wires." *Evening Star* (Washington, DC). October 19, 1888.
13. "The wire tappers." *Evening Star* (Washington, DC). October 22, 1888.
14. "Robbed the pool sellers." *St. Paul Globe*. July 16, 1889.
15. "Got off easy." *St. Paul Globe*. July 18, 1889.
16. "Tapping wires." *Helena* (MT) *Independent*. July 28, 1889.
17. "Tapped the wire successfully." *Sun* (New York). November 18, 1889.
18. "Tapped the wires." *St. Paul Globe*. February 23, 1890.
19. "Wire tappers at work." *Pittsburg Dispatch*. June 1, 1890.
20. "Jumped to certain death." *St. Paul Globe*. June 4, 1890.
21. "Tapped the wires." *San Francisco Call*. June 24, 1890.
22. "Fallon convicted." *San Francisco Call*. August 28, 1890.
23. "Fallon pardoned." *San Francisco Call*. December 4, 1892.
24. "Indicted for wire tapping." *Evening Star* (Washington, DC). September 1, 1890; "A surprised wire tapper." *Evening World* (New York). January 3, 1891.

25. "They were chumps." *St. Paul Globe*. September 19, 1890; "Nipped in the bud." *Evening Star* (Washington, DC). September 19, 1890.
26. "Tapped the wires." *St. Paul Globe*. December 18, 1890.
27. "A shrewd swindle." *St. Paul Globe*. January 6, 1892.
28. "That pool-room swindle." *Sun* (New York). January 8, 1892.
29. "Wire tappers at work." *Sun* (New York). March 27, 1892.
30. Ibid.
31. "Accused of tapping wires." *Evening World* (New York), May 18, 1892.
32. "War on pool rooms." *St. Paul Globe*. October 7, 1892.
33. "Pool room wires." *Evening Star* (Washington, DC). November 5, 1892.
34. Ibid.
35. "Poolroom sales." *San Francisco Call*. January 15, 1893.
36. "The poolroom nuisance." *New York Tribune*. January 16, 1893.
37. "Wire-tapper Martin again." *Sun* (New York). January 30, 1893.
38. "The king of the wire-tappers." *Sun* (New York). March 12, 1893.
39. "Arrest of supposed wire tappers." *Sun* (New York), May 3, 1893.
40. "State News." *Breckenridge News* (Cloverport, KY). June 28, 1893.
41. "Wires tapped." *St. Paul Globe*. March 30, 1894.
42. "Caught tapping wires." *Sun* (New York). April 15, 1894.
43. "Live topics about town." *Sun* (New York). April 17, 1894.
44. "Western wire-tappers known here." *New York Tribune*. August 9, 1894; "McCloskey goes free." *Evening Bulletin* (Maysville, KY). August 15, 1894.
45. "Wire tappers in a boat." *Washington Times*. September 14, 1894.
46. "Crisp sporting comment." *Washington Times*. September 17, 1894.
47. "Wire-tapper number three." *Washington Times*. September 24, 1894.
48. "Can't reach the wire tappers." *Washington Times*. September 22, 1894.
49. "Wire-tappers." *Washington Times*. January 6, 1895.
50. "Unlawful to tap wires." *Washington Times*. February 17, 1895.
51. "Six poolrooms in the Tenderloin." *New York Tribune*. November 21, 1894.
52. "Tapped the wires." *Wichita Eagle*. January 10, 1896.
53. "Pool rooms hit." *St. Paul Globe*. March 15, 1896.
54. "Bookmakers are badly swindled." *San Francisco Call*. March 15, 1896.
55. "Poolrooms lost many thousands." *San Francisco Call*. March 16, 1896.
56. "Was the wire tapped." *Evening Times* (Washington, DC). March 16, 1896.
57. "Arrest of a wire-tapper." *San Francisco Call*. March 20, 1896.
58. "Tapping the bookies bar'l." *Omaha Daily Bee*. April 4, 1896.
59. Ibid.
60. "Wire tappers caught." *Kansas City Journal*. June 4, 1897.
61. "$500,000 from wire tapping." *Sun* (New York). June 5, 1897.
62. "Just lies, say the police." *Sun* (New York). June 6, 1897.
63. Ibid.
64. "In the field of sports." *Evening Times* (Washington, DC). November 24, 1897.
65. "Wire tappers traced." *Richmond* (VA) *Dispatch*. December 23, 1900.
66. "Pool rooms hit for over $20,000." *Richmond* (VA) *Dispatch*. February 8, 1902.
67. "Utah men tapped the wires." *Salt Lake Herald*. February 27, 1902.
68. "Wire-tappers rob poolroom." *San Francisco Call*. September 14, 1902.
69. "Alleged wire-tapper caught by detectives." *San Francisco Call*. September 18, 1902.

70. "The poolrooms beaten." *New York Tribune*. January 6, 1903.
71. "Must not swindle swindlers." *Daily Journal* (Salem, OR). November 3, 1903.
72. "Made some big killings." *Minneapolis Journal*. November 4, 1903.
73. "In the underworld." *Salt Lake Tribune*. June 3, 1906.
74. "Wire tappers in Windsor loop." *Minneapolis Journal*. July 6, 1906.
75. "Bet on the races after getting results." *San Francisco Call*. January 12, 1907.
76. "Wireless to poolrooms." *New York Tribune*. August 30, 1907.
77. "Santa Rosa poolroom is defrauded." *San Francisco Call*. April 3, 1908.
78. "Jailed before he had committed any crime." *Salt Lake Herald*. April 20, 1908; "Wire tapper released." *San Francisco Call*. April 22, 1908.
79. "Wire tappers caught at work." *San Francisco Call*. April 23, 1908.
80. "Wire tappers work Tacoma." *Evening Statesman* (Walla Walla, WA), May 19, 1908.
81. "Jockey Club wars on poolrooms." *Washington Herald*. June 26, 1908.
82. "College men tap telegraph wires." *San Francisco Call*. January 22, 1909.
83. "Wire tappers freed; no crime, says judge." *San Francisco Call*. June 4, 1909.
84. "Live wire worth seventy thousand." *Salt Lake Tribune*. November 11, 1909.
85. "Million dollars is the clean-up of wire tappers." *Washington Herald*. April 11, 1910; "Wire tapping job starts inquiry." *San Francisco Call*. April 12, 1910.
86. "Wire-tappers make big haul." *Times* (VA) *Dispatch*. February 26, 1911.

Chapter 5

1. "Deciphering secret messages." *Milan* (TN) *Exchange*. May 2, 1878.
2. "Wire tapping in Chicago." *New York Tribune*. October 18, 1883.
3. "Sharp wire tappers." *Omaha Daily Bee*. March 25, 1887.
4. "Tapping wires." *Helena* (MT) *Independent*. July 28, 1889.
5. "Bucket shop wires tapped." *St. Paul Globe*, May 10, 1893.
6. "Telephone wire tapped." *Sun* (New York). March 23, 1898.
7. "Cotton men in a panic." *New York Times*. September 30, 1899.
8. Ibid.
9. "The false cotton quotations." *Sun* (New York). October 1, 1899.
10. "The panic in cotton." *New York Tribune*. October 1, 1899.
11. "Very nicely hidden." *Minneapolis Journal*. June 21, 1902.
12. "Very hard to get at." *Minneapolis Journal*. June 23, 1902.
13. "Chamber folk find the leak." *Minneapolis Journal*. October 7, 1902.
14. "Charged with wiretapping." *New York Tribune*. January 4, 1906.
15. "To steal quotations." *Deseret Evening News* (Salt Lake City). July 20, 1906.
16. "Bucketshops getting the knockout blows." *Marion* (OH) *Daily Mirror*. June 3, 1908.
17. *Arizona Republican*. June 4, 1908.
18. "Allege grain quotation frauds." *New York Tribune*. May 16, 1909.
19. "Paid money for market quotes." *Mahoning Dispatch* (Canfield, OH). May 21, 1909.
20. "Wire-tapping case." *Deseret Evening News* (Salt Lake City). September 13, 1909.
21. "Charged with wire tapping." *Sun* (New York). September 8, 1909.
22. "End of the bucket shop." *Washington Herald*. April 3, 1910.

23. "End of the bucket shop." *Washington Herald.* April 3, 1910.
24. Ibid.
25. "More bucket shops." *Washington Herald.* April 4, 1910.
26. "Tracing Marrin service." *New York Tribune.* May 4, 1910.
27. "Cuts off bucket shops." *New York Tribune.* July 8, 1910.
28. "Prominent Salt Lake broker and associates are arrested on charges of wiretapping." *Salt Lake Tribune.* September 2, 1911.
29. Ibid.
30. "Aftermath of wire tapping." *Salt Lake Tribune.* September 5, 1911.
31. "New complaint in wire-tapping case." *Salt Lake Tribune.* September 29, 1911.
32. $50,000 damages sought by Lowe." *Salt Lake Tribune.* February 22, 1913.
33. "Wire-tapping by telephone is charge against Warnock." *El Paso Herald.* January 15, 1914.
34. "New stock scheme arouses Wall St." *New York Tribune.* March 16, 1914.

Chapter 6

1. "Pool sellers swindled." *New York Tribune.* October 14, 1883.
2. "Caught in the trap." *San Francisco Call.* January 15, 1895.
3. Ibid.
4. Ibid.
5. *Evening Star* (Washington, DC). January 25, 1895.
6. "Electric stealing." *Evening Star* (Washington, DC). February 16, 1895.
7. Ibid.
8. Ibid.
9. Ibid.
10. "The story of a fake." *Sun* (New York). October 20, 1895.
11. "Arrested for wire tapping." *Sun* (New York). October 10, 1896.
12. "Wire tapping detected." *Wall Street Journal.* October 10, 1896.
13. "Tapping telegraph wires." *San Francisco Call.* November 21, 1896.
14. "Wire tapper sentenced." *Deseret Evening News* (Salt Lake City). May 24, 1898.
15. "There are purchasable spies in many households." *San Francisco Call.* January 1, 1899.
16. Ibid.
17. Ibid.
18. "Betrayal of a trust." *San Francisco Call.* January 2, 1899.
19. "Do nothing but listen." *Salt Lake Herald.* March 30, 1903.
20. "Tap wire says Amory." *New York Tribune.* May 15, 1903.
21. "Attack on Kilburn." *New York Tribune.* August 11, 1903.
22. "Kilburn denies plot." *New York Tribune.* August 12, 1903.
23. "A unique decision." *Paducah* (KY) *Sun.* December 23, 1903.
24. "Wire tapping suit." *Sun* (New York). February 3, 1905.

Chapter 7

1. "Telegrams to the Star." *Evening Star* (Washington, DC). July 20, 1877.
2. "An operator's joke and its result." *Salt Lake Evening Democrat.* April 5, 1886; "A remarkable plot." *Los Angeles Herald.* April 6, 1886.

3. "The strike." *Springfield* (OH) *Globe Republic*. April 7, 1886.
4. "A quiet day." *National Republican* (Washington, DC). April 7, 1886.
5. "Wire-tapping conspiracy." *New York Tribune*. April 17, 1886.
6. "Conspiracies and conspirators." *Omaha Daily Bee*. April 19, 1886.
7. "That wire-tapping." *Dodge City* (KS) *Times*. April 22, 1886.
8. "Martin Irons' troubles." *St. Paul Globe*. September 23, 1886.
9. "Martin Irons acquitted." *Washington Critic* (DC). February 28, 1888.
10. "Powderly makes reply." *Omaha Daily Bee*. October 5, 1889.
11. "Wires tapped." *Los Angeles Herald*. July 12, 1894.
12. "Wardner is now an armed camp." *Salt Lake Herald*. April 27, 1899.
13. "Admits tapping of union wires." *New York Times*. June 8, 1916.
14. "Swann orders inquiry on new complaint of wire-tapping abuses." *Evening World* (New York). June 13, 1916.
15. "Union men demand wire tapping list." *Sun* (New York). June 13, 1916.
16. "Gotham police tap union wires." *Labor World* (Duluth, MN). June 24, 1916.
17. "Wire-tap upheld by magistrate." *New York Tribune*. July 31, 1916.
18. "Wire-tapping approved." *New York Tribune*. August 1, 1916.
19. "Hist, old sleuth caught after he taps union wire." *Labor World* (Duluth, MN). August 30, 1919.

Chapter 8

1. "A detective story." *Iola* (KS) *Register*. February 12, 1886.
2. "Unique robbery." *El Paso Daily Herald*. August 18, 1900.
3. "Arrested the swindlers." *Phillipsburg* (KS) *Herald*. October 18, 1900.
4. "Brigands go across the border." *San Francisco Call*. September 12, 1904.
5. "Wire tapping scheme." *Deseret Evening News* (Salt Lake City). September 22, 1905.
6. *New Ulm* (MN) *Weekly Review*. June 6, 1888.
7. "High school boy taps wireless at his home." *Los Angeles Herald*. May 29, 1907.
8. *St. Paul Globe*. January 18, 1889.
9. "This and that." *Evening World* (New York). February 21, 1889.
10. "The Trowbridge-Ingersoll scandal settled by divorce." *Sun* (New York). February 21, 1889.
11. "Ingersoll resigns, but not in person." *Evening World* (New York). March 2, 1889; "A wedding." *New York Tribune*. January 10, 1891.
12. "New phase of the Ingersoll-Trowbridge scandal." *Sun* (New York). October 9, 1892.
13. "Learns wife loves chauffeur; urges her to marry him." *Tacoma Times*. November 4, 1911.
14. "Sister-in-law uses periscope on wife in divorce tangle." *New York Tribune*. March 11, 1920.

Chapter 9

1. "Washington." *Public Ledger* (Memphis, TN). May 30, 1868.
2. "The Arkansas imbroglio." *Sun* (New York). April 20, 1874.
3. "The government keyhole." *National Republican* (Washington, DC). May 20, 1881.

4. *Weekly Democratic Statesman* (Austin, TX). June 2, 1881.
5. "The Connecticut wire tapping law." *Sun* (New York). October 6, 1891.
6. "Ohio Legislature." *News-Herald* (Hillsboro, OH). March 2, 1893.
7. "Laid over." *Salt Lake Herald*. February 21, 1894.
8. "Kept tally on Boyce." *Washington Post*. January 22, 1898.
9. "The Ohio boodle inquiry." *Guthrie* (OK) *Daily Leader*. January 23, 1898.
10. "Take six alleged policy men." *New York Tribune*. September 4, 1908.
11. "Raid poolroom centre." *New York Tribune*. May 1, 1909.
12. "Wire-tapper's heavy sentence." *New York Tribune*. October 19, 1909.
13. "Try private detective." *New York Tribune*. January 5, 1912.
14. "Sensational developments in wire tapping case." *Seattle Star*. November 16, 1912.
15. "Candidate for council guilty of misdemeanor." *San Francisco Call*. February 16, 1913; "Free Nordskog." *Seattle Star*. December 1, 1913.
16. "Sulzer lived in a net of spies." *Burlington* (VT) *Weekly Free Press*. October 23, 1913.
17. "Bill to prevent tapping of phones now in Senate." *New York Tribune*. April 19, 1916.
18. "Mayor authorized phone spy on priest." *New York Times*. April 19, 1916.
19. Ibid.
20. "Wiretappers shut from charity row." *Sun* (New York). April 21, 1916.
21. "Grand jury takes up telephone tapping." *Sun* (New York). May 5, 1916.
22. "Hatchet buried, Thompson won't fight Whitman." *Evening World* (New York). May 15, 1916.
23. "Police in two years tapped 350 phones." *New York Times*. May 17, 1916.
24. "Mitchel calls on Thompson demanding that police head be put on stand at once." *Evening World* (New York). May 18, 1916.
25. Ibid.
26. "Tapping of wires defended by Woods." *Sun* (New York). May 18, 1916.
27. "Telephone tapping." *Evening World* (New York). May 19, 1916.
28. "2 indictments due in Farrell phone tapping." *New York Tribune*. May 21, 1916.
29. Ibid.
30. "Phone tapping revelations involve the foreign and Mexican situations." *Democratic Banner* (Mt. Vernon, OH). May 23, 1916.
31. "Swann turns inquiry to the employer of Burns." *Sun* (New York). May 25, 1916; "How detectaphone was put in room." *Sun* (New York). May 20, 1916.
32. "Phone tap inquiry to broaden to-day." *Sun* (New York). May 26, 1916.
33. "Reassure the public." *Evening World* (New York). May 27, 1916.
34. "$5,000 honorarium charges denounced; wire tapping inquiry ended." *Sun* (New York). May 27, 1916.
35. "Hint at tap information that may interest U.S." *Sun* (New York). May 27, 1916.
36. "Dr. Potter falls dead in office." *Sun* (New York). August 19, 1916.
37. "Priests victors over Mitchel." *New York Tribune*. September 16, 1916.
38. "Kingsbury faces trial again. *Sun* (New York). April 30, 1917.
39. "Woods testifies at opening of wiretapping trial." *New York Tribune*. May 22, 1917.
40. "Judge orders acquittal in wire-tapping case." *Evening World* (New York). May 24, 1917.

41. "Court upholds tapping of wires." *New York Times*. June 3, 1916.
42. "Garbage wiretap charged." *New York Times*. June 3, 1916.
43. "Wire tapping lands bottle." *Tacoma Times*. August 25, 1916.
44. "Bill to check wiretapping." *New York Tribune*. February 16, 1917.
45. "Outlining charges against Thompson." *Sun* (New York). September 7, 1917.
46. "Wire tapping laid to police captain." *Sun* (New York). October 9, 1917.
47. "Police captain named by Swann as wire tapper." *New York Tribune*. October 10, 1917.
48. "Capt. Falconer reinstated." *New York Tribune*. February 26, 1918.
49. "Telephone tapped, charges Corrigan." *Sun* (New York). November 3, 1917.
50. "Telephone tapping bill too lenient." *Sun* (New York). February 18, 1918.
51. "Governor follows wishes of Wilson on all war bills." *New York Tribune*. April 25, 1918.
52. "Penalties for wire-tapping." *Washington Times*. October 25, 1918.
53. "Fickert to charge U.S. investigator with wire-tapping." *New York Tribune*. November 30, 1918.
54. "Shades of Tom Mooney." *Labor World* (Duluth, MN). January 18, 1919.
55. "Gave an unfair trial." *Ogden Standard*. July 23, 1919; "A federal dictograph listened in for Mooney." *New York Tribune*. August 3, 1919.
56. "Brothers of politician are accused of bookmaking." *New York Tribune*. March 16, 1919.
57. "Court bars wire evidence." *Sun* (New York). March 29, 1919.
58. "Graft grand jury shadowed, also fears wire tapping." *New York Tribune*. November 9, 1919; "Grand Jury moves to shut off leaks." *Sun* (New York). November 28, 1919.
59. "Porter ousted as Enright aid on vice charge." *New York Tribune*. March 21, 1920; "Policeman tells of finding Porter in woman's flat." *Evening World* (New York). April 20, 1920.
60. "$1,000 per day on moonshine." *Washington Herald*. June 18, 1920.
61. "Tap phone wire; trap 4 for fraud." *New York Times*. November 6, 1920.
62. "Steinbrink named to assist O'Brien in city's inquiry." *Evening World* (New York). November 23, 1920.
63. "Wire tapping charged in insurance inquiry." *Washington Post*. November 8, 1921.

Chapter 10

1. "Miller Reese Hutchinson." Wikipedia.com, accessed October 10, 2013.
2. "Dictograph carries sound." *Washington Herald*. December 7, 1906.
3. "Over the wire." *Brownsville Daily Herald*. December 11, 1906.
4. "Dictograph in operation." *Evening Star* (Washington, DC). April 21, 1907.
5. "Progressive events in the field of electricity." *Omaha Daily Bee*. November 3, 1907.
6. "Device that magnifies sound." *Bismarck Daily Tribune*. February 26, 1910.
7. Francis Phillips. "Wall Street's new way of trapping financial deceivers." *Deseret Evening News* (Salt Lake City). June 18, 1910.
8. Ibid.
9. "Jackpots their favorite game." *Marion* (OH) *Daily Mirror*. May 3, 1911.
10. "Bribery in Legislature." *News-Herald* (Hillsboro, OH). May 4, 1911.

11. "The dictograph will tattle every word you say." *Tacoma Times*. May 22, 1911.
12. "Diegel trial is full of ginger." *Marion* (OH) *Daily Mirror*. June 23, 1911.
13. "Dictagraph a witness at the trial of members of the Ohio Legislature indicted for bribery." *Rock Island* (IL) *Argus*. June 28, 1911; "Dictograph as a detective." *Libby* (MT) *Herald*. September 7, 1911.
14. "Dictagraph will stay." *Washington Times*. February 23, 1912.
15. "Inventor of dictograph will testify in trial of indicted Ohio legislator." *Rock Island* (IL) *Argus*. March 8, 1912.
16. "Dictagraph gathers in evidence otherwise impossible to obtain." *El Paso Herald*. March 9, 1912.
17. "Senator Cetone found guilty." *Democratic Banner* (Mt. Vernon, OH). December 6, 1912.
18. "Corruption charge against officials." *Salt Lake Tribune*. September 9, 1911.
19. "Charge Mayor with bribery." *Marion* (OH) *Daily Mirror*. September 9, 1911; "Mayor under investigation." *Hickman* (KY) *Courier*. October 19, 1911.
20. "Dictagraph." *Kentucky Irish American* (Louisville). February 24, 1912.
21. "Dictagraph records faked." *New York Tribune*. February 19, 1912.
22. "Unearth Democratic plan to buy Kentucky for McCreary ticket." *Hartford* (KY) *Republican*. November 3, 1911.
23. "McNamara Brothers." Wikipedia.com, accessed October, 2013.
24. "Woman stenographer who played an important role in bribery inquiry during McNamara dynamiting case." *Bismarck Daily Tribune*. December 8, 1911; "The dreadful dictograph." *San Francisco Call*. January 9, 1912.
25. "Detective Burns and his mechanical detector." *Bisbee* (AZ) *Daily Review*. May 21, 1912.
26. "Dictagraph caught labor president." *Hawaiian Star* (Honolulu). February 18, 1912.
27. "That walls in reality have ears discovered." *Bisbee* (AZ) *Daily Review*. February 18, 1912; "McNamara's chum let government put in dictagraph." *Evening World* (New York). October 9, 1912.
28. "Conversations of ironworkers' president." *Rock Island* (IL) *Argus*. February 22, 1912; "Dictagraph gathers in evidence otherwise impossible to obtain." *El Paso Herald*. March 9, 1912.
29. "Stormy scene occurred in bribery case." *Arizona Republican*. February 1, 1912.
30. "Say Darrow was trapped by dictograph." *Daily Capital Journal* (Salem, OR). March 13, 1912.
31. "Dictagraph used in the Darrow case." *Salt Lake Tribune*. March 14, 1912.
32. "Pictures tell story." *Rock Island* (IL) *Argus*. April 1, 1912.
33. "Confession taken on a dictagraph." *San Francisco Call*. February 2, 1912.
34. "Dictagraph report was largely faked." *Labor World* (Duluth, MN). February 24, 1912.
35. "If dictograph gets into general use." *Tacoma Times*. February 10, 1912.
36. "Eight hundred mechanical detectives working in Chicago." *Hawaiian Star* (Honolulu). February 15, 1912.
37. "Enter the dictograph; so whisper it gently." *Washington Herald*. February 18, 1912.
38. "Dictagraph used in divorce case." *Albuquerque Evening Herald*. February 23, 1912.

39. "New terror to evildoers is that scientific eavesdropper, the dictagraph." *New York Tribune*. February 25, 1912.

40. Ibid.

41. "Dictagraphs put in bandits' cells record secrets." *Evening World* (New York). March 4, 1912.

42. "Dictograph, the modern detective." *Caucasian* (Shreveport, LA). April 7, 1912.

43. "The invisible detective." *Daily Capitol Journal* (Salem, OR). April 18, 1912.

44. "What her husband heard through the dictagraph." *El Paso Herald*. May 7, 1912.

45. "Agent of publicity." *Public Ledger* (Maysville, KY). June 4, 1912.

46. "Scientific eavesdropping." *Literary Digest*. June 15, 1912, pp. 1239–1240.

47. "Gov. Blease in on graft for pardon." *Bismarck Daily Tribune*. July 17, 1912; "Dictagraph records." *Pickens* (SC) *Sentinel*. July 25, 1912.

48. "13 Detroit aldermen arrested for boodling." *Jasper* (IN) *Weekly Courier*. August 2, 1912.

49. "Caught in the act." *Day Book* (Chicago). October 2, 1912.

50. "Dictagraph evidence convicts man of murder." *Omaha Daily Bee*. October 27, 1912.

51. "Burns men watching illegal registration." *Sun* (New York). November 3, 1912.

52. "Dictagraphs used to prevent frauds." *Salt Lake Tribune*. November 11, 1912.

53. "Dictagraph in graft case." *Sun* (New York). December 13, 1912.

54. "Deposed rector trapped in study by a dictagraph." *Evening World* (New York). December 28, 1912.

55. "Rector who received women visitors in his bedroom." *Seattle Star*. January 9, 1913.

56. "Dictagraph rigged up in court to give jury proof that it worked." *Evening World* (New York). January 24, 1913.

57. "Dictagraph used in blackmail case." *Washington Times*. January 25, 1913.

58. "Inventor fears use by crooks of his dictagraph." *Evening World* (New York). January 29, 1913.

59. "Fortune teller is convicted by the dictagraph." *Evening World* (New York). February 1, 1913; "Guilt is proved by dictagraph." *Times Dispatch* (Richmond, VA). February 2, 1913.

60. "Legislators caught in dictagraph trap." *Sun* (New York). February 12, 1913.

61. "Spying on legislators." *Forest Republican* (Tionesta, PA). March 26, 1913.

62. "The inculcation of virtue by machinery." *San Francisco Call*. March 29, 1913.

63. "Trailed for two years; held for murder." *Washington Times*. April 4, 1913.

64. "Word recording machine." *Watchman and Southron* (Sumter, SC). April 16, 1913.

65. "Dictagraph owner and Burns at odds." *Sun* (New York). May 4, 1913.

66. "Detectaphone traps many." *Sun* (New York). April 15, 1913; "U.S. has torpedo with brain as a surprise, says inventor." *Washington Times*. February 10, 1917.

67. "Osage officers under arrest." *Tulsa Daily World*. May 22, 1913.

68. "Chiefs discuss dictagraph use." *Washington Times*. June 12, 1913.

69. "Only something akin to miracle can avert strike." *Times Dispatch* (Richmond, VA). July 13, 1913.

70. "Dictagraph traps alleged swindlers." *Sun* (New York). August 13, 1913.
71. "Guilty plea offered in Orner case." *El Paso Herald*. October 10, 1913.
72. "Stilwell boss talks to be made evidence." *New York Tribune*. October 30, 1913.
73. "Famous detective and his wife guarded because of death threats." *Bismarck Daily Tribune*. November 6, 1913.
74. "Detectagraph used by state board is undoing of Dr. Card." *San Francisco Call*. November 27, 1913.
75. "Dictograph in local hotel nabs crook." *Tacoma Times*. January 24, 1914.
76. "Demonstrates dictograph at National Press Club." *Washington Times*. February 22, 1914.
77. "Caught with dictagraph." *Hopkinsville Kentuckian*. February 26, 1914.
78. "Dictagraph employed in strike inquiry." *Washington Herald*. January 12, 1914.
79. "Detecta-phone used on auto official." *Sun* (New York). March 8, 1914.
80. "Damages asked by Mrs. Gerlinger." *Daily Capital Journal* (Salem, OR). April 4, 1914.
81. "Dictagraph in murder office." *Washington Herald*. July 2, 1914.
82. "Wife's diary bares secrets of all Dr. Carman's patients." *Evening World* (New York). July 2, 1914.
83. "Mrs. Carman relates the story of her trials." *New York Tribune*. May 23, 1915.
84. "'Ware of dictograph." *Bourbon News* (Paris, KY). November 27, 1914.
85. "Legislator would bar dictographs." *Washington Times*. January 24, 1915.
86. "Court rejects dictagraph." *Labor Advocate* (Cincinnati). December 4, 1915.
87. "Privacy as a right." *Public Ledger* (Maysville, KY). May 9, 1916.
88. "Car strike spies used dictagraph." *New York Tribune*. August 13, 1916.
89. "Use beauty to blackmail rich." *El Paso Herald*. September 18, 1916.
90. "Dictagraph mystery stirs Bremerton row." *Tacoma Times*. March 23, 1917.
91. "Dictagraph users held by magistrate for listening in." *New York Tribune*. August 29, 1919.
92. "Misusing the dictagraph." *New York Tribune*. September 1, 1919.
93. "U.S. agents seized in huge graft plot on city liquor men." *Sun* (New York). October 23, 1919.
94. "Use dictagraph to trap Dansey case suspect." *Washington Times*. December 13, 1919.
95. "Hammonton police force is dismissed." *Evening Public Ledger* (Philadelphia). December 13, 1919.
96. "Court restrains officers from use of dictograph." *El Paso Herald*. May 21, 1920.
97. "Courting." *Bismarck Tribune*. October 15, 1921.
98. "Late New York theatrical gossip." *Salt Lake Tribune*. August 8, 1909; "Belasco—The Only Law." *Washington Times*. October 3, 1909.
99. Advertisement, Orpheum. *Bismarck Daily Tribune*. August 1, 1912.
100. William J. Burns. "How I caught the land grafters." *Salt Lake Tribune*. October 27, 1912; "Dictograph owner and Burns at odds." *Sun* (New York). May 4, 1913.
101. "Modern Maud." *Broad Ax* (Salt Lake City). December 28, 1912.
102. "Eavesdropping in a play." *New York Tribune*. January 19, 1913; "When the criminal hears his own voice speak the words that doom him." *Day Book* (Chicago). June 12, 1913.

Bibliography

"Accused of tapping wires." *Evening World* (New York). May 18, 1892.
"Admits tapping of union wires." *New York Times*. June 8, 1916.
Advertisement, Orpheum. *Bismarck Daily Tribune*. August 1, 1912.
"Aftermath of wire tapping." *Salt Lake Tribune*. September 5, 1911.
"Agent of publicity." *Public Ledger* (Maysville, KY). June 4, 1912.
"Allege grain quotation frauds." *New York Tribune*. May 16, 1909.
"Alleged electricity thief." *St. Paul Globe*. July 21, 1898.
"Alleged wire-tapper caught by detectives." *San Francisco Call*. September 18, 1902.
Arizona Republican. June 4, 1908.
"The Arkansas imbroglio." *Sun* (New York). April 20, 1874.
"Arraign sleuth as wiretapper." *Seattle Star*. November 27, 1912.
"Arrest of a wire-tapper." *San Francisco Call*. March 20, 1896.
"Arrest of supposed wire tappers." *Sun* (New York). May 3, 1893.
"Arrested for wire tapping." *Sun* (New York). October 10, 1896.
"Arrested the swindlers." *Phillipsburg* (KS) *Herald*. October 18, 1900.
"Attack on Kilburn." *New York Tribune*. August 11, 1903.
"Badly cinched." *Sacramento Record-Union*. October 19, 1888.
"Belasco—the only law." *Washington Times*. October 3, 1909.
"Bet on the races after getting results." *San Francisco Call*. January 12, 1907.
"Betrayal of a trust." *San Francisco Call*. January 2, 1899.
"Bill to check wiretapping." *New York Tribune*. February 16, 1917.
"Bill to prevent tapping of phones now in Senate." *New York Tribune*. April 19, 1916.
"Bookmakers are badly swindled." *San Francisco Call*. March 15, 1896.
"A boy spy in Dixie." *National Tribune* (Washington, DC). May 31, 1888.
"Bribery in legislature." *News-Herald* (Hillsboro, OH). May 4, 1911.
"Brigands go across the border." *San Francisco Call*. September 12, 1904.
"Brothers of politician are accused of bookmaking." *New York Tribune*. March 16, 1919.
"Bucket shop wires tapped." *St. Paul Globe*. May 10, 1893.
"Bucketshops getting the knockout blows." *Marion Daily Mirror* (OH). June 3, 1908.
"Burns men watching illegal registration." *Sun* (New York). November 3, 1912.
Burns, W. J. "How I caught the land grafters." *Salt Lake Tribune*. October 27, 1912.
"Can electricity be stolen." *Sacramento Record-Union*. May 5, 1897.

"Candidate for council guilty of misdemeanor." *San Francisco Call*. February 16, 1913.
"Can't reach the wire tappers." *Washington Times*. September 22, 1894.
"Capt. Falconer reinstated." *New York Tribune*. February 26, 1918.
"Car strike spies used dictagraph." *New York Tribune*. August 13, 1916.
"Caught in the act." *Day Book* (Chicago). October 2, 1912.
"Caught in the trap." *San Francisco Call*. January 15, 1895.
"Caught tapping wires." *Sun* (New York). April 15, 1894.
"Caught with dictagraph." *Hopkinsville Kentuckian*. February 26, 1914.
"Chamber folk find the leak." *Minneapolis Journal*. October 7, 1902.
"Charge mayor with bribery." Marion (OH) *Daily Mirror*. September 9, 1911.
"Charged with wire tapping." *Sun* (New York). September 8, 1909.
"Charged with wiretapping." *New York Tribune*. January 4, 1906.
"Chiefs discuss dictagraph use." *Washington Times*. June 12, 1913.
"Claim he stole light current." *San Francisco Call*. March 28, 1903.
"College men tap telegraph wires." *San Francisco Call*. January 22, 1909.
"Confession taken on a dictagraph." *San Francisco Call*. February 2, 1912.
"The Connecticut wire tapping law." *Sun* (New York). October 6, 1891.
"Conspiracies and conspirators." *Omaha Daily Bee*. April 19, 1886.
"Conversations of ironworkers' president." *Rock Island* (IL) *Argus*. February 22, 1912.
"Corruption charge against officials." *Salt Lake Tribune*. September 9, 1911.
"Cotton men in a panic." *New York Times*. September 30, 1899.
"Court bars wire evidence." *Sun* (New York). March 29, 1919.
"Court rejects dictagraph." *Labor Advocate* (Cincinnati). December 4, 1915.
"Court restrains officers from use of dictograph." *El Paso Herald*. May 21, 1920.
"Court upholds tapping of wires." *New York Times*. June 3, 1916.
"Courting." *Bismarck Tribune*. October 15, 1921.
"Crisp sporting comment." *Washington Times*. September 17, 1894.
"Cuts off bucket shops." *New York Tribune*. July 8, 1910.
"Damages asked by Mrs. Gerlinger." *Daily Capital Journal* (Salem, OR). April 4, 1914.
"Deciphering secret messages." Milan (TN) *Exchange*. May 2, 1878.
"Demonstrates dictagraph at National Press Club." *Washington Times*. February 22, 1914.
"Deposed rector trapped in study by a dictagraph." *Evening World* (New York). December 28, 1912.
"Detecta-phone used on auto official." *Sun* (New York). March 8, 1914.
"Detectagraph used by state board is undoing of Dr. Card." *San Francisco Call*. November 27, 1913.
"Detectaphone threat makes Burns smile." *New York Tribune*. April 22, 1916.
"Detectaphone traps many." *Sun* (New York). April 25, 1913.
"Detective Burns and his mechanical detector." Bisbee (AZ) *Daily Review*. May 21, 1912.
"A detective story." Iola (KS) *Register*. February 12, 1886.
"Device that magnifies sound." *Bismarck Daily Tribune*. February 26, 1910.
"Dictagraph." *Kentucky Irish American* (Louisville). February 24, 1912.
"Dictagraph a witness at the trial of members of the Ohio Legislature indicted for bribery." Rock Island (IL) *Argus*. June 28, 1911.
"Dictagraph employed in strike inquiry." *Washington Herald*. January 12, 1914.
"Dictagraph evidence convicts man of murder." *Omaha Daily Bee*. October 27, 1912.

"Dictagraph gathers in evidence otherwise impossible to obtain." *El Paso Herald.* March 9, 1912.
"Dictagraph in graft case." *Sun* (New York). December 13, 1912.
"Dictagraph in murder office." *Washington Herald.* July 2, 1914.
"Dictagraph mystery stirs Bremerton row." *Tacoma Times.* March 23, 1917.
"Dictagraph owner and Burns at odds." *Sun* (New York). May 4, 1913.
"Dictagraph put in bandits' cells record secrets." *Evening World* (New York). March 4, 1912.
"Dictagraph records." *Pickens* (SC) *Sentinel.* July 25, 1912.
"Dictagraph records faked." *New York Tribune.* February 19, 1912.
"Dictagraph rigged up in court to give jury proof that it worked." *Evening World* (New York). January 24, 1913.
"Dictagraph traps alleged swindlers." *Sun* (New York). August 13, 1913.
"Dictagraph used in blackmail case." *Washington Times.* January 25, 1913.
"Dictagraph used in divorce case." *Albuquerque Evening Herald.* February 23, 1912.
"Dictagraph used in the Darrow case." *Salt Lake Tribune.* March 14, 1912.
"Dictagraphs used to prevent frauds." *Salt Lake Tribune.* November 11, 1912.
"Dictagraph users held by magistrat for listening in." *New York Tribune.* August 29, 1919.
"Dictagraph will stay." *Washington Times.* February 23, 1912.
"Dictograph as a detective." *Libby* (MT) *Herald.* September 7, 1911.
"Dictograph carries sound." *Washington Herald.* December 7, 1906.
"Dictograph caught labor president." *Hawaiian Star* (Honolulu). February 17, 1912.
"Dictograph in local hotel nabs crook." *Tacoma Times.* January 24, 1914.
"Dictograph in operation." *Evening Star* (Washington, DC). April 21, 1907.
"Dictograph report was largely faked." *Labor World* (Duluth, MN). February 24, 1912.
"Dictograph, the modern detective." *Caucasian* (Shreveport, LA). April 7, 1912.
"The dictograph will tattle every word you say." *Tacoma Times.* May 22, 1911.
"Diegel trial is full of ginger." *Marion* (OH) *Daily Mirror.* June 23, 1911.
"Do nothing but listen." *Salt Lake Herald.* March 30, 1903.
"Dr. Potter falls dead in office." *Sun* (New York). August 19, 1916.
"The dreadful dictograph." *San Francisco Call.* January 9, 1912.
"Eavesdropping in a play." *New York Tribune.* January 19, 1913.
"Eight hundred mechanical detectives working in Chicago." *Hawaiian Star* (Honolulu). February 15, 1912.
"Electric light wire is tapped." *San Francisco Call.* December 11, 1901.
"Electric stealing." *Evening Star* (Washington, DC). February 16, 1895.
"Electric wire tapped." *Scranton Tribune.* January 7, 1898.
"Electricity is property." *Evening Times* (Washington, DC). April 29, 1897.
"End of the bucket shop." *Washington Herald.* April 3, 1910.
"Enter the dictograph; so whisper it gently." *Washington Herald.* February 18, 1912.
Evening Bulletin (Maysville, KY). August 1, 1892.
Evening Star (Washington, DC). January 25, 1895.
"Fallon convicted." *San Francisco Call.* August 28, 1890.
"Fallon pardoned." *San Francisco Call.* December 4, 1892.
"The false cotton quotations." *Sun* (New York). October 1, 1899.
"Famous detective and his wife guarded because of death threats." *Bismarck Daily Tribune.* November 6, 1913.

"A federal dictograph listened in for Mooney." *New York Tribune*. August 3, 1919.
"Ferry Cafe. 16 Market Street." *San Francisco Call*. April 19, 1903.
"Fickert to charge U.S. investigator with wire-tapping." *New York Tribune*. November 30, 1918.
"The field of electricity." *Omaha Daily Bee*. February 24, 1898.
"The field telegraph." *Wichita Eagle*. June 12, 1887.
"$50,000 damages sought by Lowe." *Salt Lake Tribune*. February 22, 1913.
"Finds druggist Beck is guilty." *San Francisco Call*. January 3, 1902.
"Fined for tapping the electric wires." *San Francisco Call*. January 5, 1902.
"$500,000 from wire tapping." *Sun* (New York). June 5, 1897.
"$5,000 honorarium charges denounced; wire tapping inquiry ended." *Sun* (New York). May 27, 1916.
"For stealing electricity." *Salt Lake Herald*. December 29, 1905.
"Fortune teller is convicted by the dictagraph." *Evening World* (New York). February 1, 1913.
"Free Nordskog." *Seattle Star*. December 1, 1913.
"Garbage wiretap charged." *New York Times*. June 3, 1916.
"Gave an unfair trial." *Ogden* (UT) *Standard*. July 23, 1919.
"General brevities." *Iola* (KS) *Register*. January 3, 1879.
"Got off easy." *St. Paul Globe*. July 18, 1889.
"Gotham police tap union wires." *Labor World* (Duluth, MN). June 24, 1916.
"Gov. Blease in on graft for pardon." *Bismarck Daily Tribune*. July 17, 1912.
"The government keyhole." *National Republican* (Washington, DC). May 20, 1881.
"Governor follows wishes of Wilson on all war bills." *New York Tribune*. April 25, 1918.
"Graft grand jury shadowed, also fears wire tapping." *New York Tribune*. November 9, 1919.
"Grand jury moves to shut off leaks." *Sun* (New York). November 28, 1919.
"Grand jury takes up telephone tapping." *Sun* (New York). May 5, 1916.
"Guilt is proved by dictagraph." *Times Dispatch* (Richmond, VA). February 2, 1913.
"Guilty plea offered in Orner case." *El Paso Herald*. October 10, 1913.
"Hammonton police force is dismissed." *Evening Public Ledger* (Philadelphia). December 13, 1919.
"Hatchet buried, Thompson won't fight Whitman." *Evening World* (New York). May 15, 1916.
"Held for tapping Edison Co's wires." *New York Tribune*. August 16, 1913.
"High school boy taps wireless at his home." *Los Angeles Herald*. May 29, 1907.
"Hint at tap information that may interest U.S." *Sun* (New York). May 27, 1916.
"Hist, old sleuth caught after he taps union wire." *Labor World* (Duluth, MN). August 30, 1919.
"How detectaphone was put in room." *Sun* (New York). May 20, 1916.
"If dictograph gets into general use." *Tacoma Times*. February 10, 1912.
"In the field of sports." *Evening Times* (Washington, DC). November 24, 1897.
"In the underworld." *Salt Lake Tribune*. June 3, 1906.
"The inculcation of virtue by machinery." *San Francisco Call*. March 29, 1913.
"Indicted for wire tapping." *Evening Star* (Washington, DC). September 1, 1890.
"Ingersoll resigns, but not in person." *Evening World* (New York). March 2, 1889.
"Inventor fears use by crooks of his dictagraph." *Evening World* (New York). January 29, 1913.

"Inventor of dictograph will testify in trial of indicted Ohio legislator." *Rock Island* (IL) *Argus*. March 8, 1912.
"The invisible detective." *Daily Capital Journal* (Salem, OR). April 18, 1912.
"Is the telephone a failure?" *Stark County Democrat* (Canton, OH). July 20, 1877.
"Jackpots their favorite game." *Marion* (OH) *Daily Mirror*. May 3, 1911.
"Jailed before he had committed any crime." *Salt Lake Herald*. April 20, 1908.
"Jersey lawmakers adjourn." *Sun* (New York). April 1, 1897.
"Jockey Club wears on poolrooms." *Washington Herald*. June 26, 1908.
"Judge orders acquittal in wire-tapping case." *Evening World* (New York). September 7, 1917.
"Jumped to certain death." *St. Paul Globe*. June 4, 1890.
"Just lies, say the police." *Sun* (New York). June 6, 1897.
"Keep tally on Boyce." *Washington Post*. January 22, 1898.
"Kilburn denies plot." *New York Tribune*. August 12, 1903.
Kilmer, George L. "Bold John Morgan." *Princeton* (MN) *Union*. January 2, 1896.
"The king of the wire-tappers." *Sun* (New York). March 12, 1893.
"Kingsbury faces trial again." *Sun* (New York). April 30, 1917.
"Laid over." *Salt Lake Herald*. February 21, 1894.
Las Vegas Free Press. July 14, 1892.
"Late New York theatrical gossip." *Salt Lake Tribune*. August 8, 1909.
"Latest news." *Edgefield* (SC) *Advertiser*. September 16, 1863.
"Learns wife loves chauffeur; urges her to marry him." *Tacoma Times*. November 4, 1911.
"Legislator would bar dictographs." *Washington Times*. January 24, 1915.
"Legislators caught in dictograph trap." *Sun* (New York). February 12, 1913.
"Live topics about town." *Sun* (New York). April 17, 1894.
"Live wire worth seventy thousand." *Salt Lake Tribune*. November 11, 1909.
"Made some big killings. *Minneapolis Journal*. November 4, 1903.
"Martin Irons acquitted." *Washington* (DC) *Critic*. February 28, 1888.
"Martin Irons' troubles." *St. Paul Globe*. September 23, 1886.
"Mayor authorized phone spy on priest." *New York Times*. April 19, 1916.
"Mayor under investigation." *Hickman* (KY) *Courier*. October 19, 1911.
"McCloskey goes free." *Evening Bulletin* (Maysville, KY). August 15, 1894.
"McNamara Brothers." Wikipedia.com, accessed October, 2013.
"McNamara's chum let government put in dictograph." *Evening World* (New York). October 9, 1912.
"Message in fragments." *Sun* (New York). March 23, 1883.
"Miller Reese Hutchinson." Wikipedia.com, accessed September 2013.
"Million dollars is the clean-up of wire tappers." *Washington Herald*. April 11, 1910.
"Misusing the dictograph." *New York Tribune*. September 1, 1919.
"Mitchel calls on Thompson demanding that police head be put on stand at once." *Evening World* (New York). May 18, 1916.
"Modern Maud." *Broad Ax* (Salt Lake City). December 28, 1912.
"More bucket shops." *Washington Herald*. April 4, 1910.
"Morgan's raid." *Wichita Eagle*. July 29, 1888.
"Mrs. Carman relates the story of her trials." *New York Tribune*. May 23, 1915.
"Must not swindle swindlers." *Daily Journal* (Salem, OR). November 3, 1903.
"New complaint in wire-tapping case." *Salt Lake Tribune*. September 29, 1911.
"New phase of the Ingersoll-Trowbridge scandal." *Sun* (New York). October 9, 1892.

"New stock scheme arouses Wall St." *New York Tribune*. March 16, 1914.
"New terror to evildoers is that scientific eavesdropper, the dictagraph." *New York Tribune*. February 25, 1912.
New Ulm (MN) *Weekly Review*. June 6, 1888.
"Nipped in the bud." *Evening Star* (Washington, DC). September 19, 1890.
"No dictagraph appears." *Sun* (New York). September 24, 1913.
"Not caught." *Salt Lake Herald*. October 16, 1883.
"The Ohio boodle inquiry." *Guthrie* (OK) *Daily Leader*. January 23, 1898.
"Ohio Legislature." *News-Herald* (Hillsboro, OH). March 2, 1893.
"$1,000 per day on moonshine." *Washington Herald*. June 18, 1920.
"Only something akin to miracle can avert strike." *Times Dispatch* (Richmond, VA). July 13, 1913.
"An operator's joke and its result." *Salt Lake Evening Democrat*. April 5, 1886.
"Osage officers under arrest." *Tulsa Daily World*. May 22, 1913.
"Outlining charges against Thompson." *Sun* (New York). September 7, 1917.
"Over the wire." *Brownsville* (TX) *Daily Herald*. December 11, 1906.
"Paid money for market quotes." *Mahoning Dispatch* (Canfield, OH). May 21, 1909.
"The panic in cotton." *New York Tribune*. October 1, 1899.
"Penalties for wire-tapping." *Washington Times*. October 25, 1918.
Phillips, Francis. "Wall Street's new way of trapping financial deceivers." *Deseret Evening News* (Salt Lake City). June 18, 1910.
"Phone tap inquiry to broaden today." *Sun* (New York). May 26, 1916.
"Phone tapping revelations involve the foreign and Mexican situations." *Democratic Banner* (Mt. Vernon, OH). May 23, 1916.
"Pictures tell story." *Rock Island* (IL) *Argus*. April 1, 1912.
"Pirates of the wires." *Juniata Sentinel and Republican* (Mifflintown, PA). July 25, 1883.
"Police captain named by Swann as wire tapper." *New York Tribune*. October 10, 1917.
"Police in two years tapped 350 phones." *New York Times*. May 17, 1916.
"Policeman tells of finding Porter in woman's flat." *Evening World* (New York). April 20, 1920.
"The pool mystery discovered." *Los Angeles Herald*. October 19, 1883.
"Pool room wires." *Evening Star* (Washington, DC). November 5, 1892.
"Pool rooms hit." *St. Paul Globe*. March 15, 1896.
"Pool rooms hit for over $20,000." *Richmond* (VA) *Dispatch*. February 8, 1902.
"Pool sellers swindled." *New York Tribune*. October 14, 1883.
"The poolroom nuisance." *New York Tribune*. January 16, 1893.
"Poolroom sales." *San Francisco Call*. January 15, 1893.
"The poolrooms beaten." *New York Tribune*. January 6, 1903.
"Poolrooms lost many thousands." *San Francisco Call*. March 16, 1896.
"Porter ousted as Enright aid on vice charge." *New York Tribune*. March 21, 1920.
"Powderly makes reply." *Omaha Daily Bee*. October 5, 1889.
"A power that cannot be stolen." *Forest Republican* (Tionesta, PA). March 17, 1897.
"Pranks of telephones." *Princeton* (MN) *Union*. July 8, 1880.
"Priests victors over Mitchel." *New York Tribune*. September 16, 1916.
"The printing telegraph." *Richmond* (VA) *Dispatch*. April 29, 1886.
"Privacy as a right." *Public Ledger* (Maysville, KY). May 9, 1916.
"Progressive events in the field of electricity." *Omaha Daily Bee*. November 3, 1907.
"Prominent Salt Lake broker and associates are arrested on charges of wire-tapping." *Salt Lake Tribune*. September 2, 1911.

"Puts a thief on wire." *Salt Lake Herald*. October 27, 1909.
"A quiet day." *National Republican* (Washington, DC). April 7, 1886.
"Raid poolroom centre." *New York Tribune*. May 1, 1909.
"Reassure the public." *Evening World* (New York). May 27, 1916.
"Rector who received women visitors in his bedroom." *Seattle Star*. January 9, 1913.
"A remarkable plot." *Los Angeles Herald*. April 6, 1886.
"A revolution in telegraphy." *Evening Times* (Washington, DC). December 27, 1898.
"Robbed the pool sellers." *St. Paul Globe*. July 16, 1889.
"Santa Rosa poolroom is defrauded." *San Francisco Call*. April 3, 1908.
"Say Darrow was trapped by dictograph." *Daily Capital Journal* (Salem, OR). March 13, 1912.
"Scientific eavesdropping." *Literary Digest*. June 15, 1912, pp. 1239–1240.
"Senator Cetone found guilty." *Democratic Banner* (Mt. Vernon, OH). December 6, 1912.
"Sensational developments in wire tapping case." *Seattle Star*. November 16, 1912.
"Shades of Tom Mooney." *Labor World* (Duluth, MN). January 18, 1919.
"Sharp wire tappers." *Omaha Daily Bee*. March 25, 1887.
"A shrewd swindle." *St. Paul Globe*. January 6, 1892.
"Sister-in-law uses periscope on wife in divorce tangle." *New York Tribune*. March 11, 1920.
"Six poolrooms in the Tenderloin." *New York Tribune*. November 21, 1894.
Smoky Hill and Republican Union (Junction City, KS). September 6, 1862.
"Sound sent by wire." *New York Tribune*. March 31, 1877.
"South Omaha man fined for tapping light wire." *Omaha Daily Bee*. August 20, 1911.
"Spying on legislators." *Forest Republican* (Tionesta, PA). March 26, 1913.
St. Paul Globe. January 18, 1889.
"State news." *Breckenridge News* (Cloverport, KY). June 28, 1893.
"Stealing electricity." *Sun* (New York). July 20, 1887.
"Steinbrink named to assist O'Brien in city's inquiry." *Evening World* (New York). November 23, 1920.
"Stilwell boss talks to be made evidence." *New York Tribune*. October 30, 1913.
"Stormy scene occurred in bribery case." *Arizona Republican*. February 1, 1912.
"The story of a fake." *Sun* (New York). October 20, 1895.
"The strike." *Springfield* (OH) *Globe Republic*. April 7, 1886.
Suburban Citizen (Washington, DC). May 26, 1900.
"Sulzer lived in a net of spies." *Burlington* (Vermont) *Weekly Free Press*. October 23, 1913.
"A surprised wire tapper." *Evening World* (New York). January 3, 1891.
"Swann orders inquiry on new complaint of wire-tapping abuses." *Evening World* (New York). June 13, 1916.
"Swann turns inquiry to the employer of Burns." *Sun* (New York). May 25, 1916.
"Swindling by telegraph." *New York Tribune*. October 17, 1883.
"Systematic methods." *St. Paul Globe*. September 8, 1900.
"Take six alleged policy men." *New York Tribune*. September 4, 1908.
"Tap phone wire; trap 4 for fraud." *New York Times*. November 6, 1920.
"Tap wire says Amory." *New York Tribune*. May 15, 1903.
"Tapped the wire successfully." *Sun* (New York). November 18, 1889.
"Tapped the wires." *New York Tribune*. April 3, 1910.
"Tapped the wires." *St. Paul Globe*. February 23, 1890.

"Tapped the wires." *St. Paul Globe*. December 18, 1890.
"Tapped the wires." *San Francisco Call*. June 24, 1890.
"Tapped the wires." *San Francisco Call*. June 29, 1895.
"Tapped the wires." *Wichita Eagle*. January 10, 1896.
"Tapping news conduit." *Sun* (New York). July 8, 1883.
"Tapping of wires defended by Woods." *Sun* (New York). May 18, 1916.
"Tapping telegraph wires." *San Francisco Call*. November 21, 1896.
"Tapping the bookies bar'l." *Omaha Daily Bee*. April 4, 1896.
"Tapping the poolsellers' wire." *Sun* (New York). October 18, 1883.
"Tapping the wires." *Daily Dispatch* (Richmond, VA). July 8, 1883.
"Tapping the wires." *Evening Star* (Washington, DC). October 19, 1888.
"Tapping wires." *Helena* (MT) *Independent*. July 28, 1889.
"Telegrams to the Star." *Evening Star* (Washington, DC). July 20, 1877.
"Telegraphic secrets." *New Ulm* (MN) *Weekly Review*. October 25, 1882.
"Telephone tapped, charges Corrigan." *Sun* (New York). November 3, 1917.
"Telephone tapping." *Evening World* (New York). May 19, 1916.
"Telephone tapping bill too lenient." *Sun* (New York). February 18, 1918,
"Telephone wire tapped." *Sun* (New York). March 23, 1898.
"That pool-room swindle." *Sun* (New York). January 8, 1892.
"That walls in reality have ears discovered." *Bisbee* (AZ) *Daily Review*. February 18, 1912.
"That wire-tapping." *Dodge City* (KS) *Times*. April 22, 1886.
"There are purchasable spies in many households." *San Francisco Call*. January 1, 1899.
"They were chumps." *St. Paul Globe*. September 19, 1890.
"13 Detroit aldermen arrested for boodling." *Jasper* (IN) *Weekly Courier*. August 2, 1912.
"This and that." *Evening World* (New York). February 21, 1889.
"To steal quotations." *Deseret Evening News* (Salt Lake City). July 20, 1906.
"Tracing Marrin service." *New York Tribune*. May 4, 1910.
"Trailed for two years; held for murder." *Washington Times*. April 4, 1913.
"The Trowbridge-Ingersoll scandal settled by divorce." *Sun* (New York). February 21, 1889.
"Try private detective." *New York Tribune*. January 5, 1912.
"2 indictments due in Farrell phone tapping." *New York Tribune*. May 21, 1916.
"Unearth Democratic plan to buy Kentucky for McCreary ticket." *Hartford* (KY) *Republican*. November 3, 1911.
"Union men demand wire tapping list." *Sun* (New York). June 13, 1916.
"A unique decision." *Paducah* (KY) *Sun*. December 23, 1903.
"Unique robbery." *El Paso Daily Herald*. August 18, 1900.
"Unlawful to tap wires." *Washington Times*. February 17, 1895.
"U.S. agents seized in huge graft plot on city liquor men." *Sun* (New York). October 23, 1919.
"U.S. has torpedo with brain as a surprise, says inventor." *Washington Times*. February 10, 1917.
"Use beauty to blackmail rich." *El Paso Herald*. September 18, 1916.
"Use dictagraph to trap Dansey case suspect." *Washington Times*. December 13, 1919.
"Utah men tapped the wires." *Salt Lake Herald*. February 27, 1902.
"Very hard to get at." *Minneapolis Journal*. June 23, 1902.

"Very nicely hidden." *Minneapolis Journal*. June 21, 1902.
"Wants to protect telephone talks." *New York Times*. March 22, 1914.
"War on pool rooms." *St. Paul Globe*. October 7, 1892.
"Wardner is now an armed camp." *Salt Lake Herald*. April 27, 1899.
"'Ware of dictograph." *Bourbon News* (Paris, KY). November 27, 1914.
"Was the wire tapped." *Evening Times* (Washington, DC). March 16, 1896.
"Washington." *Public Ledger* (Memphis, TN). May 30, 1868.
"A wedding." *New York Tribune*. January 10, 1891.
Weekly Democratic Statesman (Austin, TX). June 2, 1881.
"Western wire-tappers known here." *New York Tribune*. August 9, 1894.
"What her husband heard through the dictograph." *El Paso Herald*. May 7, 1912.
"When the criminal hears his own voice speak the words that doom him." *Day Book* (Chicago). June 12, 1913.
"Wife's diary bares secrets of all Dr. Carman's patients." *Evening World* (New York). July 2, 1914.
"Wire-tap upheld by magistrate." *New York Tribune*. July 31, 1916.
"Wire-tapper Martin again." *Sun* (New York). January 30, 1893.
"Wire-tapper number three." *Washington Times*. September 24, 1894.
"Wire tapper released." *San Francisco Call*. April 22, 1908.
"Wire tapper sentenced." *Deseret Evening News* (Salt Lake City). May 24, 1898.
"Wire-tapper's heavy sentence." *New York Tribune*. October 19, 1909.
"The wire tappers." *Evening Star* (Washington, DC). October 22, 1888.
"Wire-tappers." *Washington Times*. January 6, 1895.
"Wire tappers at work." *Pittsburg Dispatch*. June 1, 1890.
"Wire tappers at work." *Sun* (New York). March 27, 1892.
"Wire tappers caught." *Kansas City Journal*. June 4, 1897.
"Wire tappers caught at work." *San Francisco Call*. April 23, 1908.
"Wire tappers freed; no crime, says judge." *San Francisco Call*. June 4, 1909.
"Wire tappers in a boat." *Washington Times*. September 14, 1894.
"Wire tappers in Windsor loop." *Minneapolis Journal*. July 6, 1906.
"Wire-tappers make big haul." *Times Dispatch* (Richmond, VA). February 26, 1911.
"Wire-tappers rob poolroom." *San Francisco Call*. September 14, 1902.
"Wire tappers traced." *Richmond* (VA) *Dispatch*. December 23, 1900.
"Wire tappers work Tacoma." *Evening Statesman* (Walla Walla, WA). May 19, 1908.
"Wire-tapping." *New Ulm* (MN) *Weekly Review*. April 2, 1884.
"Wire-tapping approved." *New York Tribune*. August 1, 1916.
"Wire-tapping by telephone is charge against Warnock." *El Paso Herald*. January 15, 1914.
"Wire-tapping case." *Deseret Evening News* (Salt Lake City). September 13, 1909.
"Wire-tapping charged in insurance inquiry." *Washington Post*. November 8, 1921.
"Wire-tapping conspiracy." *New York Tribune*. April 17, 1886.
"Wire tapping detected." *Wall Street Journal*. October 10, 1896.
"Wire tapping in Chicago." *New York Tribune*. October 18, 1883.
"Wire tapping job starts inquiry." *San Francisco Call*. April 1910.
"Wire tapping laid to police captain." *Sun* (New York). October 9, 1917.
"Wire tapping lands bottle." *Tacoma Times*. August 25, 1916.
"Wire tapping scheme." *Deseret Evening News* (Salt Lake City). September 22, 1905.
"Wire tapping suit." *Sun* (New York). February 3, 1905.
"The wire was tapped." *Washington Post*. October 19, 1888.

"Wireless telegraphic codes." *New York Tribune*. October 10, 1907.
"Wireless to poolrooms." *New York Tribune*. August 30, 1907.
"Wires tapped." *Los Angeles Herald*. July 12, 1894.
"Wires tapped." *St. Paul Globe*. March 30, 1894.
"Wires tapped on Emeryville races." *San Francisco Call*. March 21, 1909.
"Wiretappers shut from charity row." *Sun* (New York). April 21, 1916.
"Woman stenographer who played an important role in bribery inquiry during McNamara dynamiting case." *Bismarck Daily Tribune*. December 8, 1911.
"Woods testifies at opening of wiretapping trial." *New York Tribune*. May 22, 1917.
"Word recording machine." *Watchman and Southron* (Sumter, SC). April 26, 1913.
"W. T. Conway is convicted." *Salt Lake Herald*. January 4, 1906.

Index

Acousticon 142
Adams, Charles H. 62–63
adultery 110
Akoulophon 142
aldermen, Detroit 172
alienation of affections 111
Alkorn, William 64–65
Altberger, John P. 74
Amalgamated Clothing Workers of
   America 103
Ambler, John 175
American Federation of Labor 104
Amory, William N. 95
Andrews, L.R. 151–152
Anti-Garbage League 133
anti-war protests 134
The Argyle Case 190–191
Arzner, Louis 16
assassination story, false 88–89
Associated Press 82–85
authority to tap 120

Babcock, Palmer B. 49
Badger Brothers 77–78
Bailey, Louise 185
Baird, Frank 46
Baltimore, Harry 181
Baltimore and Ohio Telegraph Company
   87–88
bank swindle 108–109
banks 145–146
Barto, Isaac N. 73
Barton, Fred 44
Baxter, Elisha 112–113
Beck, Ignatz 15–16
Bentley, G.B. 176–177
Bethel, Frank H. 120

Bethel, Union N. 23
Bingham, George 107–108
Birmingham, AL 65–66
black voters 154
Blease, Coleman 171–172
blind tigers 172
Boardman, William 128
boardwalk construction 174
Boissonnault, Gaston 185
book, fiction 58
bookie joints 24–63
bookies, illegality of 24
bootleggers 180, 188–189
Bowman, J.M. 17–18
Boyce, H.H. 114–115
Boyle, Frank 39
Brady, Peter J. 103–106
breach of promise 184–185
break-in, William Burns 126
Bremerton Washington 187
Brewer, W.S. 83–85
bribery 148–153, 172, 176–177
Bridgers, W.W. 181
Brooks, J.C. 108
Brooks, Joseph 112–113
Brooks-Baxter Affair 112–113
Brown, Mary 133
Brown, Sam 156
bucket shops 64–81; background 65;
   campaigns against 74–77; decline of
   72; operations of 75; and Western
   Union 71
bugging of companies by themselves
   144–145
Bunker Hill mine 103
Burdick, E.A. 189
Burlingame, Alvah W. 118–119, 134

Burns bugged LA prison cells 157
Burns Detective Agency 147–148, 155–160, 162–163, 173, 184, 189; prison bugged by 182; tapped 117
Burns, William J. 125, 147–151, 161, 171, 177–179; acting as police 155; and films 190; and stage play 190–191
Burtis, Al 29
Butler, Benjamin 112
Butler, J.T. 157
Butler, William S. 123

California Penal Code 35–36
California State Board of Medical Examiners 182
Cameron, Ed 44
Canadian Pacific Railway 109
capital punishment, and dictograph 173
Card, W.S. 182
Carman, Edwin 185
Carman, Florence 185
Carney, Thomas 67–68
Cetone, George K. 152
Chandler, William 85
Charities Department 121
charities scandal, NYC 118–132
Chattanooga, TN 10
Chicago Board of Trade 44, 65, 70, 73, 79
Christie Grain Company 69
ciphers 29, 40, 46, 48–49, 102
Civil War, U.S. 5–11
Clark, Charles W. 56
Clement, Alice 189–190
Cleveland, Grover 88–89
Clifford, Russell 12–13
Clinton, Hugh 18
codes, telegraph 6
Cohalan, Daniel F. 133
Cohen, Seth 16
Coley, Clarence T. 127
collaboration, police, corps, private detectives 124–126
Collins, Edward 31
Collins, J.W. 48
Columbia Pool Room, St. Paul 31–32
Columbus Ohio 47–48, 114–115
commercial electricity theft 13–18
commercial users, of bugs 144–145
commodity exchanges 64–81
common carrier doctrine 24, 71
Common Council Committee, Detroit 172
Common Council of Atlantic City, NJ 174

companies, self-bugging 144–146
company hype, published as news 156
complaints, not made 24–26
Coney Island, NY 29
Conkling, Roscoe 113
Conlon, Thomas J. 116
Connecticut, law against 113
connections, electric, illegal 13–14
conspiracy, employer 98–102
Content, Walter 67
convictions, difficulties 13
Conway, W.T. 17
Cope, M. 180
corporations, self-bugging 184
Corrigan, Joseph 135–136
corruption, political 171–172
Cotton, Joe 46
cotton exchanges 65–66
Coughlan, A.C. 98–102
court decisions 70, 73, 96, 117
Craig, H. 173
crime detecting, and tapping 124–125
crimes, personal 107–109
crytograms 21

Dannenberg Detective Agency 106
Darrow, Clarence 156–160
Dean, T.B. 152–154
Deford, Gus 171–172
Deneen, Charles 146
Densmore, John B. 136–137
Department of Charities, NYC 119
detectaphone, Burns name for dictograph 180
detective agencies, private 104–106
Detroit aldermen 172
Devanney, Bernard J. 116
dictagraph (variant spelling) *see* dictographs
Dictograph Products Company 142
Dictograph Products Corporation 188
dictographs 118; as bug, first mention 144–145; evidence, admitted 174; evil possibilities 168; explained 162–163; explained by Turner 176; first major case 147–152; injunction against 189; misuse of 188; praised 156–157, 168–169, 169–171; rentals 163, 166, 185; tested and verified in court 175
Diegel, Rodney J. 151–152
Diffenbacher, T.C. 61
Diffendorf, Theodore 40–41
Dio, Fannie 175–176

divorce cases 110–111, 134–135, 165, 169
Donaldson, Thomas B. 139–140
Downey, Peter 43
Downing, Guy 182
Drew, Walter 159
Dudley, Thomas P. 38
Duff, Rath 176
Dugan, Jack 46
Dunham, B.V. 109
Dunn, Thomas J. 67
duplex line 53
duplex systems 87
duplicitous nature of financial firms 145
Dykeman, Conrad F. 135
dynamiting campaign 155

eavesdropping 23, 187–188
Eberhart, Adolph 146
Edmunds, Frank 37
Edwards, William Seymour 176–177
E.E. Hutton & Company 78
Egan, Martin 127, 130
election dispute 112–113
electricity: home 12; stealing 12–18
Elliott, Walter 67
Ellsworth, George 5–6, 8, 10–11
Enright, Richard 138
Europe 166
execution 8
Exposed by the Dictograph 190
The Exposure of the Land Swindlers 190
express company swindle 107–109
extortion 180, 187
extradition 155

faked transcripts 161
Falconer, John 135
Fallabom, Eugene 169
Fallabom, Marguerite 169
Fallon, William 34–36
false news: dispatches 83–84; stories 88–89
false transcriptions 152
Farnum, A.H. 70
Farrell, May 111
Farrell, W.B. 119–127
fast wire service 76
federal agents: bugging federal agents 188–189; tappers 187
Felder, Thomas B. 172
Ferry Café 16
Fickert, Charles M. 136–137
fictional representation 58

financial scams 145–146
fingerprinting 55
fire insurance adjusting 139–140
Flynn, William J. 188–189
formal request, to tap 120
Foster, Robert J. 159
Frank, Lloyd 184–185
Frankfort, KY 11
Franklin, Bert H. 156
frauds 139
Furlong, Thomas 98–102, 107–108
Fust, Harry 29

Gallatin, TN 8
galvanometer 35
gambling establishments, horses 24–63
gangs 50–51, 58
Garfield, James A. 113
Garland, Marion 187
Gary, Indiana 152–154
Geddis, Walter 49
General Acoustic Company 142, 185
Gerlinger, Gertrude 184–185
Germany 13–14
Giachono, John 172
Gibbons, John H. 96–97
Gibson, Walter 153–154
Giles, M.E. 187
Gill, Frank 60
Gleason's Custom House (NYC) 33
Glinnan, Thomas 172
Gompers, Samuel 156
Gould, Jay 98–102
Gould System secret service 98–102
governor, NYS, tapped 117–118
Graham, Frank A. 89
Graham, John 39
grain quotations 73
Grant, Ulysses S 7–8, 112–113
Grierson, Benjamin 8
Guttenburg track 29, 38

hackers 109–110
Haggerty, Harry W. 140
Hamele, Ottamar 180
Hammond, T.B. 61
Hanford, Cornelius 117
Hanna, Mark 114–115
Harrington, John R. 159
Harris, John 40–41
hearing aid, electronic 142–143
heliograph 20–21
Helmus, Carlos 189

Hennessy, John A. 181
high school student 110
Himmelblau, Meyer 153–154
Hitchcock, Eula 156
Hockin, Herbert S. 155
Hocking Valley Crash 145
Hodgson, John 18
Holmes, Carroll 16
Honest Ballot Association 173
Hooker, Joseph 8
Hopper, Albert 60
House of Delegates, West Virginia 176–177
how to tap bookies 25–28
Howard, George Bronson 190
Hoxie, H.M. 98–102
Humstone, W.C. 26, 38
Hunters Point (NY) 28
Hutchinson, Miller Reese 142
Hylan, John 138

Idaho 103
Illinois Central Railroad 169
impeachment 112, 117–118
Independent Electric Light and Power Company 16–17
Independent Telegraph Company 89–90
information sharing 104–106
Ingersoll, Clark Jonathan 110–111
Insurance Commission of Pennsylvania, as tapper 140
insurance fraud 181
intangible items 13–14
intercom systems 142–144
International Association of Bridge and Structural Iron Workers 154–160
International Association of Chiefs of Police 180
International Ladies' Garment Workers' Union 103, 106
international plot, Morgan bankers 126
Iron Workers 154–160; bugged 156
Irons, Martin 98–102

Jackson, George M. 100–102
Jacobson, Jack 187–188
jail cells bugged 183–184, 189
jail time 90
Jaynes, Frank 36
Jeffries, Frank B. 37
Jerome Park 28
Johnson, Andrew 112
Johnson, D.S. 154

Johnson, William 37
Johnson Detective Agency 154
Jones, Edith 189
J.P. Morgan & Company 123–126
judicial review, whitewash 106
jurisdictional disputes 103–104

Kansas City Board of Trade 69–70
Kentucky politics scandal 154
Kerbey, Joseph Orton 8
kidnapping 155
Kilmer, George 10
King, Charles E. 180
King, George 116–117
Kingsbury, John A. 121
Kingsley, C.H. 34
Knapp, Clyde D. 184
Knights of Labor 98–102
Knotts, Thomas E. 152–154
Kupfer, Ruth Baker 111
Kupfer, Walker F. 111

labor unions 154–160; bugged 181, 184, 186–187
larceny 17–18; grand 49
laws: against tapping 14; Connecticut 113; defective 47; deficient 49, 85; electricity 17; intangibles 13; NYS 118–119; Ohio 113; Utah 114
lawyers bugged, by state 156
Leach, George T. 18
Leach, John 139
Lee, Robert E. 7
Leehan, William J. 177–178
legal actions, on bookies 29
legal decisions, telegraph responsibilities 29
legal rulings 49–50, 63, 133, 151
legal system 12
legislation: New York State 136; proposed 85, 134, 186
Lehman, C.H. 188
Levinsky, Leo 133
Lexington, KY 11
Liberty Bonds 134
lighting, retail 15–16
linemen, availability of 25
Literary Digest 171
Little Rock, Arkansas 112–113
Llewellyn Iron Works 155
Lockwood Committee 139
Loft, George W. 126
Long Island City 25

Lord, Frank 129
Lorimer, William 161
Los Angeles 57
Los Angeles Police Department jail cells bugged 156
Los Angeles Times bombing case 154–160
Louisville 44
Lowden, Frank 134
Lowe, DeWitt B. 77–78
Lowell, MA 12

Maier, Frank 76
Maier, Nathan 181
Manning, Bradley (Chelsea) 5
Markham, Henry 36
Marrin, Thomas 76
Massey, William A. 173
Mattfeldt, Charles L. 57
May, E.F. 77–78
Mays Landing, New Jersey 174
McAllister, J.F. 78
McCalla, W.W. 107
McClintock, Jonas 169
McCloskey, J.G. 47
McCreary, James B. 154
McDonnell, A.G. 182
McDonnell, Charles E. 121
McGowan, C.G. 161
McKeighan, Frank 98–102
McKenna, M.J. 96–97
McManigal, Ortie 155–159
McManus, Michael 14
McNally, John 46
McNamara, James B. 154–160
McNamara, John J. 154–160
McNutt, William H. 52
Means, Gaston Means 124
media attention 26
media coverage, Ohio 149–150
Mellon, Nora McMullen 165
Mellow, Andrew W. 165
messages, false 26–28
meters, electric 15
Metropolitan Street Railway (NYC) 95
Michigan 67
milk cartel 172
Miller, Charles W. 158
Mills, Luther Laflin 83
Mills, William Woods 133
Milwaukee Electric Railway and Lighting Company 15
mining strike 103
Minneapolis 15

Minneapolis Chamber of Commerce 69–71
Minneapolis General Electric Company 15
Minneapolis Milk Company 172
Minot, T.S. 78–79
mischief, malicious 31
misdemeanor 71
Mitchel, John 103–106, 119–128
Mittleberger, J.H. 47
Mizner, Wilson 190
monitoring party lines, by telephone company 93–95
Montani, Geno 168
Mooney, Tom 136–137
moonshiners 138–139
morality 43, 166
Moran, Charles 53–55
Morgan, John Hunt 6, 10
Morse, Harry 34
Mortimer, Alfred G. 174–175
Mortimer, George 127
Morton, W.G. 61
Moss, Frank 103–106, 122
movies about 190
Mundell, W.A. 173
municipal bugging 187
munitions 126
murder case 177–178, 185
Murdock, James 172
Mutual Union News 25
Mutual Union Telegraph Company 28
Myers, Allen O. 115

Nagle, J.W. 34–36
National City Bank, self bugging 146
National Dictograph Company 156
National Erectors Association 155
National News Company 97
National Press Club 183
National Secret Telephone Company 21
National Security League 134
Nelson, C.A. 172
New England Securities Purchase and Sales Company 80–81
New Jersey, county jail bugged 189
New Jersey Legislature 14
New Orleans Cotton Exchange 68–69
New York Building-Loan Banking Company 95–96
New York City 58–59
New York City Bar Association 136
New York Edison Company 18

220 Index

New York Police Department: agreement with New York Telephone 120–124; jail cells, bugged 146–147, 168; length of time as tappers 120; as tappers 118–132, 137–138, 139, 175–176; tapping experts 116; tapping gamblers 115–116; tapping labor 103–106; tapping numbers and dates 121–122; tapping room Church St. 123; and underworld 138; wire tapping squad, examined 129–130
New York State Appellate Division 133
New York State Federation of Labor 104
New York Stock Exchange 79
New York Telephone Company 23, 120–128, 133
Newcastle Wyoming jail cell bugged 173
news agencies, as targets 82–94
Nichols, Clarence 158
Nichols, George 184
Nichols, Sam J. 171–172
Nichols, S.W. 100
Nixon, George S. 173
Nordskog, Arne A. 117
Nye, George B. 147
Nyhoff, John 153

Ocean City, New Jersey officials, entrapped 180
Oddie, Tasker 173
Offley, William M. 123
Ohio, law against 113
Ohio Legislature scandal 147–152
Ohio Supreme Court ruling 151
oil quotations 66–67
O'Leary, James 51
Omaha Electric Light and Power Company 18
The Only Law 190
operators: blames 40; surveillance of 40; suspected 50–51
operators, telegraph 8, 29, 52; blamed 69; criminal 39–40; and racing 41–42; scams 32; style of 30, 53; suspended 39
operators, telephone 23
Oppenheim, Myron 68
organized gangs 39–40
origins of electronic bugs (dictographs) 142–145
Otis, John C. 114–115
Owens, Fred J. 49
Owensboro Kentucky cell bugged 183–184

Pacific Coast Jockey Club 50
Pacific State Telephone and Telegraph Company 63–64
Pacific Telephone and Telegraph Company 90–93
Palmer & Singer Manufacturing Company 184
Pan-American News Association 83–85
pardons, prison 171–172
party lines 93–95
pay outs, bookie 31
Peck, A.E. 14–15
Pecora, Ferdinand 187–188
Pennsylvania House of Representatives, bugged 177
Pennycook, William G. 16
perjury 120
Perkins, Ray M. 77–78
pervasiveness, of dictographs 162–166, 185–186
petition for privacy 23
Phelen, Sid 65
Phillips, Francis 145–146
Pinkerton Agency 61, 77
Pinkham, C.B. 182
Pittman, Key 173
plants, by employers 155
Platt, Chester C. 118
Platt, Thomas 113
Poindexter, J.H. 58
police and private detectives, collaboration 104–105, 157
police: interrogation tactics 34; power 133; as tappers, court ruling 133
police, tapping 118–132; gamblers 115–116; labor 103–106; personal use 134–135
policewoman, advice 189–190
policy sellers (gambling) 115
political usages 112–140
politicians: bugged 147–152; entrapped by private detectives 180; non-verbal strategy 177; self-bugging 146
pool rooms 24–63; structure 42–43; war on 41, 62
Porter, Augustus Drum 138
Postal Telegraph Company 71
Potter, Daniel, death of 132
Potter, Dr. D.C. 119–132
Potter, Dean 119
Powderly, Terence 102
Prall, Anning S. 133
Pratt, John T. 184

Preparedness Day bombing 137
Preusser, Richard E. 80–81
price manipulation 68–69
privacy 19–23, 121, 171, 186; breeched 98–102; rights 90, 92; threatened 149
private and public combined attacks on labor 154–160
private detective agencies 169; bugging citizens 177–178; California, bugging partners 182; and New York Telephone 122; tapping 116–117
Prohibition 188–189
prosecutions 41
public controversy 121–124
public sentiment 84–85
Public Service Commission (NY) 23
Public Service Commission (NYS) 121
publicity 104, 161–163
Pulver, Frank 18

quadruplex systems 87

railroad strike 98–102
railroading, of labor unions 154–160
Ransom, William 136
Ray, Edgar R. 169
Rayson, Mary 37
Read, George 175
Recorder of Deeds 144
recording 141; automatic 178, 191; drawbacks 178; expense and difficulties of 142; faked 153–154; mechanical, attempts 166–168
Reed, William 155
religious leader, entrapped 174–175
renters of dictographs, vetted 166
Restmeyer, William 133
results, delayed 30
retail electricity theft 13–18
rewards posted 155
Rhea, Walter 61
Rhodes, S.U.G. 176
Richardson, B. 44
Richmond, L.L. 100
Robinson, David 95
Robinson, Joseph H. 63
Rockefeller, John D., Jr. 173
rogues' gallery photos 55
roneophone 191
Root, Elihu, Jr. 173
Rothschild, Alonzo 84
Rowland, Henry T. 22
Ruhnke, Arthur R. 172

Rutledge, Roy 183–183
Ryan, Frank M. 154–160

St. Louis 13, 33–34
Salt Lake City 17–18
Sams, E.E. 180
San Antonio 108
San Bernardino, CA 13
San Francisco 15–16, 35–36
San Francisco Call 90–93
San Francisco Examiner 90–93
San Francisco Gas and Electric Company 17
Sarasohn, H.S. 154
Saylor, E.B. 73
Schreiter, Eddie 172
Schwartz, Henry 34
Schwartz, M. 106
Scranton, PA 14
secrecy 21–23
security 19–20
Segal, Louis 111
self-bugging 163–164
senatorial election, bribery 114–115
Senes, Anthony 187–188
service cut-offs 76–77
Seymour, Frederick 126
Seymour & Seymour 123–128
Sharon, William 64
Shelton, W.T. 139
Sherman, William Tecumseh 6
Sing Sing Penitentiary, prisoner bugged 182
Smith, Bartlett 127
Smith, George Stewart 23
Smith, Hugh 110
Snodgrass, John T. 69–70
Snowden, Edward 5
Sorger, Fred 138
South Carolina Legislature 171–172
spies 8
stage plays 190–191
Standard Oil 184; self bugging 146
Stanley, Henry 72
Stearn, J.G. 108
stenographers 141, 158, 166; dual 158; records, acceptability 186
Stilwell, Stephen J. 181–182
sting operations 82–84, 180; employer 98–102
stock exchanges 64–81
Stone, Oscar M. 44, 58, 89–90
Stoneman, George 8

strikes, railroad 180–181
Stringer, Arthur 58–59
Strong, Charles 119
Strong Commission (NYS) 121
Suburban Electric Light Company 14
Sulzer, William 23, 117–118
Summerfield, H.M. 53–55
Sunset Telephone and Telegraph Company 90–93
Superior Grain Exchange 72–73
Supreme Court, U.S. 73
Swann, Edward 104–106, 120, 134
Swayze, John L. 120
Sweeney, John 46
syndicates 33, 39–40

Tacoma police, tapping 133–134
Tandlich, Samuel 175–176
tapping: ease 19–20; editorial 128
targets for 141–142
Taylor, Thomas 50
technology, changes 141–142
telegrams: confidence in 29; demise of 60–61; lines 6; multiplex 22; nationwide 5; poles 10; printing 21–22; versus telephone 19–20
telegraphers and racing 41–42
telegraphy instruments, quadruplex 38
telegraphy, wireless 22–23
telephone 19–20; acceptance of 20; boys 71; companies 90–93, 168; exchanges 12; predictions for 21
Tempel, Gustave A. 13
Texas 50
Texas Supreme Court 96
Thaw, Harry 165
Thomas, George Henry 7
Thompson, George, L. 103–106, 121–128
Thompson, Maggie 38
Thompson, William 134
Thompson Committee 125–129
Thompson Legislative Committee (NYS) 103–106
torture 156
trade unions 98–106
train robbery 109
transcripts, accuracy of 166
trespassing 85
trials 49
Trowbridge, Ann 110–111
Trowbridge, Rutherford 110–111
Turner, Kelly Monroe 142–145, 149–150, 165–166, 171, 176, 178–179; demonstrates dictograph, to journalists 183
Tyson, Fay 116–117

union busting campaigns 155
Union Express Company 107
unions, labor 98–106
United Papers 82
United States Congress members, demonstration to 143
United States Department of Justice 74–75; bugging agents 188–189; bugging itself 145; wiretapping by 74–75
United States Department of Labor 136–137; as tapper 137
United States Department of the Army 168
United States Department of the Navy 168
United States Post Office 73
United States Secret Service 136, 144–145, 166; bugging itself 145
United States Treasury Department, bugging itself 145
United States War Department 188
university student 109–110
unlawful imprisonment 156
Untermyer, Samuel 139
Utah, law against 114
Utah Light and Railway Company 17

Van Name, Calvin D. 133
Van Pelt, Harry 187–188
Van Winkle, C.F. 64–65
verbatim transcripts 141
Vines, Thomas H. 34
voice identification 152, 186
vote buying 154
voter fraud 173

Wabash Railroad 172
Waddell, James B. 116
Wade, W.K. 37
Wadham, Perry W. 36–37
Wall Street 145–146
Wall Street Journal 89
Warnock, H.V. 80
Warwick Turf Exchange 61–62
Washington, D.C. 29–31
Weihe, W.C. 161
West, Duval 189
West Virginia 98
West Virginia Legislature 186

Western Federation of Miners 184
Western Macaroni Company 17–18
Western Union Telegraph Company 24, 26, 28, 31, 32–34, 35–36, 37, 38–40, 41–42, 50–51, 57, 64–65, 71, 73, 79, 89–90, 96–97, 113; and bucket shops 71; Commercial News Bureau 68–69; indicted 74; in-house investigators 34; investigations by 46–47; race department 40; vs. Baltimore and Ohio Telegraph Company 87–88; vs. U.S. Dept of Justice 76
wheat market 71
White, Charles 189
White, Henry 116–117
White House involvement 113
Whitehead, S.B. 34
Whitman, Charles 119
Wilkie, John 145
Wilkinson's detective agency 43
Willard, William 44
Williams, L.L. 187
Willison, E.A. 180
Williston, C.A. 154
Wilson, Woodrow 137
The Wire Tappers 58–59
wireless telegraphy 60
wires, underground 87
wiretappers: absolved, by courts 132; types of 85–86
wiretapping skills 141–142
Wisconsin Grain and Stock Company 72
Wolfe Building (NYC) 18
Wood, H.A. 16–17
Woods, Arthur 103–106, 119
Woods, Elliott 143–144
World War I 134, 188

Young, J. Milton 140

www.ingramcontent.com/pod-product-compliance
Ingram Content Group UK Ltd.
Pitfield, Milton Keynes, MK11 3LW, UK
UKHW041953140426
5217IPUK00015B/787